Graphs, Orders, Infinites and Philosophy

Graphs, Orders, Infinites and Philosophy

Daniel Parrochia

iSTE

WILEY

First published 2023 in Great Britain and the United States by ISTE Ltd and John Wiley & Sons, Inc.

ISTE Ltd
27-37 St George's Road
London SW19 4EU
UK

www.iste.co.uk

John Wiley & Sons, Inc.
111 River Street
Hoboken, NJ 07030
USA

www.wiley.com

Library of Congress Control Number: 2022948075

British Library Cataloguing-in-Publication Data
A CIP record for this book is available from the British Library
ISBN 978-1-78630-897-9

Contents

Introduction

This book deals with some of the possible applications of mathematics to philosophy. In a previous work (see Parrochia 2018), we have shown how philosophers, from the birth of philosophy in Greece until now, and throughout the complex and sometimes overflowing developments in the history of philosophy, have benefited from the advanced mathematics of their time, which have often inspired them.

We now want to apply this view to today's mathematics and philosophy. It seems to us that in the great reservoir of mathematical structures, particularly those encountered in graph theory, in the theory of orders (including partial orders), as well as in a number of algebraic or geometric configurations, even infinite ones, current philosophers can find new and revolutionary modes of organization of their own thoughts.

Although the reading of the present text assumes some familiarity with mathematics, we write essentially for the philosophers, or say, for the mathematical philosophers, rather than for the specialists of the fields we mention.

Kant, in the past, considered that the risk of philosophy was to wander in nature, but that, conversely, mathematicians could not find, outside their field, either the earth which could carry them, or the water which allowed them to swim. Everything was like in Ovid's chaos: *instabilis tellus, innabilis unda* (see Kant 1922, pp. 582–583)[1].

1 Kant was exactly speaking of "the uncertain ground of pure and ever, transcendental concepts (instabilis tellus, innabilis unda) where they are neither able to stand nor to swim, taking only a few hasty steps, the vestiges of which are soon swept away" (in German: "wo der Grund (instabilis tellus, innabilis unda) ihnen weder zu stehen, noch zu schwimmen erlaubt, und sich nur flüchtige Schritte tun lassen, von denen die Zeit nicht die mindeste Spur aufbehält"). The implicit reference to Ovid refers to *Metamorphoses*, I, 15: "Utque erat et tellus illic et pontus et aer, sic erat instabilis tellus, innabilis unda, lucis egens aer". In English: "Though there was land and sea and air, it was unstable land, unswimmable water, air needing light".

In fact, mathematics is one of the main sources of intelligibility of reality. But its role does not stop there. It can also provide philosophy with more precise concepts than the ones we ordinarily use to interpret the sensible world. As we shall see, discoveries in mathematics sometimes limit or dispute the claims of past doctrines, but they may also suggest new ways of posing problems.

We first wish to illustrate these affirmations by some reflections on graph's problems, as the mathematicians today conceive them. We will go on by some considerations on order, pre-order and partial order, then carry out more geometric and topological investigations. In the end, we tackle the question of infinity and come to possible applications in our domain of some particular graphs (incomparability, fuzzy, neutrosophic or polyhedral graphs). We only use common notations and remain, the most of the time, as clear as possible, in order to give the philosophical reader some useful tools for developing their own thought.

Let us now define what a "philosophical system" is for us, i.e. the "unit" on which philosophical considerations are based. I emphasize this point: the history of thought, the sociology of knowledge may be concerned with philosophical ideas. But the proper history of philosophy is concerned with *systems*, or, at least, philosophical constructions which must explicitly explain why they are not – or cannot be – systems.

Definition: We will call a *philosophical system* a theory, expressed in the common language or a saturated extension of it, which tries to give a conceptual image of the whole world, or eventually explains why such an image is impossible to construct.

Commentary: The sentence "expressed in the common language or a saturated extension of it" allows us to distinguish between philosophy and science. Generally, philosophy is written in a language very close to the natural language and that may be read by everyone. For example, this is the case in Plato's dialogues, in the works of French authors from the 18th century such as Rousseau and Diderot, or in the essay by Fichte entitled *The destination of man*. However, in a more elaborated form, philosophy can make use of more or less abstract concepts and sometimes refers to scientific language (see, for instance, the *Logical Investigations* by Husserl). The expression "saturated extension" comes from logic and was introduced by Jean Ladrière (see Ladrière 1972)[2].

2 To keep it simple and not get into too technical considerations of model theory, we can say that a logical system is *saturated* if any addition of a supplementary thesis makes it contradictory or, if it tolerates certain forms of contradiction in itself, inconsistent with respect to a particular operation different from classical negation.

Moreover, in order to complete this distinction between science and philosophy, the term "world" has to be interpreted in a larger sense than the term "universe". Then we can separate philosophy from this branch of physics called "cosmology". Classical philosophy is a kind of existential cosmology supporting the non-physical dimensions of the world. For a classical philosopher, the universe or, a fortiori, the observed universe, is only a small part of a larger entity that the German-Swiss philosopher Karl Jaspers (1883–1969) named "the Encompassing"[3].

In this context, the classic philosophical operation consists of postulating that there exists a mapping:

$$\phi : W \to S$$

between a metaphysical entity W, which is supposed to represent the whole world in its complexity[4], and a philosophical system S, which captures its main features.

When a philosopher thinks that this operation is impossible, then they generally try to show that, in some particular system S_P written by a philosopher P, there exists some linguistic expression x to which P did not give any sense, so we cannot exhibit any corresponding concrete entity in W. In symbols:

$$\exists x \in S_P \mid \phi_P^{-1}(x) = \emptyset.$$

Commentary: Following the suggestions of Hume, Kant was probably the first philosopher who applies this method. He proves that we cannot have any knowledge of entities about which we get no information (in his language, no empirical or pure intuitions), so that the three great metaphysical "objects" named God, World and Soul have no corresponding reality in the concrete experience. Wittgenstein pursue this task in his *Tractatus logico-philosophicus* and began to hunt down all the linguistic expressions that philosophers used in a naive way and that were in the same situation. This project, in a more or less strong form, was thereafter continued by all so-called *analytical philosophy*.

3 Because of the subject–object split, which always places our consciousness outside the world of objects, being as a whole can be neither object nor subject. According to Jaspers, it must be the "encompassing" that manifests in this split. Objects as subjects therefore arise against the background of this encompassing, which remains partly obscure. It is clarified only by objects, and it becomes all the more clear as the objects are more clearly present to consciousness. But, for all that, it does not itself become an object. It is therefore basically what, through thought, only announces itself. We never meet it itself, but everything we do meet, we meet in it.

4 This complexity is assumed, but it is not yet known when the philosopher begins their work. In order to justify the possibility of this operation, some philosophers like M. Gueroult have spoken of this entity W as a "common real" (see Gueroult 1979). Others have simply postulated that there is, in everyone, a "thought of a frame of reference" (see, for example, Granier 1977).

In this book, we want to explore the actual possibility of constructing today's philosophies by studying the different forms of organizations that they can take. We will show that mathematics gives a lot of tools to build robust conceptual architectures so that we may hope that systematic philosophy is always possible[5].

This undoubtedly rests on several postulates which can be listed as follows:

– (P_1) It is possible to represent without bias, in a single form and of limited dimension – what is called a "philosophical system" – the essential of what constitutes our experience of the world.

– (P_2) A philosophical system is useful and even necessary (if not, why waste time building it?).

– (P_3) The relational structure (generally expressed as a graph[6]) of a philosophical system is connected in some sense to the relational structure of the world. But the correspondence between them is not an isomorphism – "the map is not the territory", as Korzybski said (see Korzybski 1933, p. 58) – but probably not even a homomorphism. For example, the structure of the world may be an asymmetric graph, while the structure of the system is symmetric. However, we require the system to represent, without noticeable distortion, the main features of the world of which it is supposed to be an image.

The first two postulates have a methodological character. The latter is clearly ontological.

This latter is probably the most problematic. It has been widely discussed, since the famous article by the American mathematician and philosopher Randal Roy Dipert (1997), of which we would like to say a few words here.

The first part of Dipert's article ("the shortcomings of logic and logical metaphysics") is one of the best argued refutations of analytical philosophy that can be found. The second part argues for a well-tempered "relationalism". But it is especially the third part ("structure, asymmetric graphs and Aristotle refuted") and the fourth part ("the world as asymmetric graph") which are the most interesting for our purpose[7].

5 This project must not be considered as an utopy or an old-fashioned dream. From Whitehead to Rescher, the idea of cognitive systematization has continued, as, moreover, it persists in our everyday life when we try to understand this one in all its aspects.
6 Let us just say here, to fix ideas, that it is a set of vertices connected by arcs or edges, depending on whether the graph is directed or not.
7 The reader can find the complete definition of symmetric and asymmetric graphs in section 1.8.

In the third part, after recalling the fundamental definitions of graph theory (simple graph, oriented graph, graph with loops, symmetric or asymmetric graphs, hypergraphs, etc.), Dipert then concentrates on structurally different graphs, i.e. non-isomorphic ones, the number of which grows according to a sequence that approaches asymptotically $2^{p/2}/p!$, when p is the number of their vertices. But the subset that interests him is that of non-isomorphic, asymmetric graphs and their subgraphs.

The theory of pure formal structures to which empiricity boils down is that of these non-isomorphic and asymmetric graphs, as well as their subgraphs. And the whole world is, in truth, nothing but an immense asymmetrical graph having a very particular structure, distinct from any other. Against Aristotle who, in *Categories* 7 (8a 14f–8a 37f) believes to refute relationalism, Dipert claims that asymmetric graphs exclude all kinds of monadism.

In the fourth part, two crucial questions are asked: "First, is it possible that the universe is just a formal structure, and that we can only think about the world as such a structure? Secondly, is it conceivable that the structure that is the world is just a graph-theoretical structure?" The answers to these two questions are positive.

Dipert observes that what the world reduces to is not necessarily a graph with a very large number of vertices. Medium-sized graphs (say between 40 and 100 vertices) already contain a lot of connections. For example, a graph of 40 vertices already contains 2^{40} connections, a number that approximates that of the basic components of our universe or the informational content of it, which by some estimates is roughly equivalent to the number of non-isomorphic graphs of order 40.

The link with physical "theories of everything" is as follows: physical objects, including elementary particles – which are in fact composite – do not correspond to vertices of the graph but to subgraphs. The physical microstructure is in fact a theoretical macrostructure from the graphic point of view. As for space and time, they correspond to frames or grids which are in fact families of paths (symmetrical for space, asymmetrical for time). Forces, fields and causal chains are also such paths among subgraphs that are identified as physical objects.

Naturally, the mind and its thoughts must have their place in this graph which thus includes in itself a mirror structure. As Dipert wrote:

> thoughts are about a phenomenon – refer to it – if they occupy a certain place in the world graph (in a "mind") and share certain structural features with the thought-about object: a thought, per

> impossible, is perfectly about an object just when its internal graphical structure is the same as the object's, and when the object occupies a location in the system of all objects that is like the thought's location in the system of (that mind's) thought (Dipert 1997, p. 357).

We might think then that, in such a theory, there is no room for minds, consciousness and other mental phenomena, unless, precisely, everything is mental, which finally, Dipert, following in this Leibniz, if not Spinoza, does not entirely exclude, because the vertices of the graph could, after all, only be "feelings".

This remarkable article surely unleashed the hostility of the philosophers who prefer Aristotelian matter to formal structures and the sentimental illusion of human warmth to the objective relationships depicted in this cold graph. But the strongest objection, developed in a short article by Oderberg (2011, pp. 6–9), was based on a mathematical argument that Shackel (2011, p. 11) very well summarized as follows:

1) suppose the world is a graph;

2) if it is a graph, it is an asymmetric graph;

3) any asymmetric graph can be turned into symmetric graph by the removal of edges;

4) the loss of less than all the edges of the world (specifically, the loss of all those not part of some symmetric subgraph, a loss consequent on certain nodes going out of existence) metaphysically entails the non-existence of the entire world (1, 2, 3);

5) Therefore, the world is not a graph (1, 4, reductio ad absurdum).

In fact, this mathematical argument is quite specious: why imagine that the asymmetric graph of the world could become symmetrical? As Shackel writes, "it's impossible for the world graph to be other than it is and hence any change at all entails non-existence" (Shackel 2011, p. 13), so Oderberg's objection is not a valid argument: "it is impossible to get rid of the nodes and edges of a world and the argument to the absurd inexistence is blocked" (Shackel 2011, p. 14).

The fact remains that the necessity of Dipert's structure may seem troublesome. However, reducing the properties to simple potentials, as Bird could have done in another model (see Bird 2007), would not remove the problem according to Oderberg. Because if there is only potential and no actualization, then nothing can really manifest itself, as he also explains in another article (see Oderberg 2012). But, as Shackel shows, there are potency graphs for which a diffusion process gradually updates some (or all) of the potentials. The phenomenon is a "snowball" and we can demonstrate that, in this type of graph (snowflake graphs), we can always choose a node from which all the potentials can be actualized.

In conclusion, Dipert's graph is perfectly plausible and this possible reduction of the world to a graph makes it possible, as well, to somehow ontologically base the attempts of philosophers over time. If it is rejected, however, it cannot be denied that these attempts can be methodologically described using the formalism of graph theory. So we do not absolutely need the assumption P_3 to justify what follows.

However, if we accept to consider the world as a graph, then, if the world is finite, there is a good chance that its graph is asymmetric. Indeed, the proportion of graphs on n vertices with non-trivial automorphisms tends to zero when n grows, which informally means that almost all finite graphs are asymmetric. In contrast, almost all infinite graphs are symmetric and, more specifically, countable infinite random graphs in the Erdös–Rényi model are, with probability 1, isomorphic to the highly symmetric Rado graph (see Erdös and Szekeres 1935).

In this text, using the language of graphs, orders and sometimes infinites, we try to represent mathematically this entity that Jaspers called "the encompassing", and which he thought it was, by definition, impossible to objectify.

In fact, mathematics has often taught us that it is possible to obtain good characterizations of an object without having to bring it in any exteriority. As Gauss showed in the case of surfaces (theorema egregium), it is very remarkable that the curvature of a geometrical object can be described intrinsically, i.e. without any reference to a "space of embedding" in which the considered object would be located. For example, the fact that an ordinary sphere is a surface with constant positive curvature is completely independent of the fact that we usually see this sphere as being immersed in our three-dimensional Euclidean space. The curvature of this sphere could very well be measured by two-dimensional intelligent beings living on the sphere (kinds of "two-dimensional ants"), from measurements of lengths and angles carried out on the sphere.

We must imagine that we are, with respect to the encompassing, in the same situation as two-dimensional ants with respect to the sphere. In principle, nothing prevents us from being able to describe it if we have the necessary information.

1

Graphs

This chapter is an introduction to our subject. After a few words about the history of graphs, we explain what we now call a "graph" in mathematics and introduce the main definitions useful to understand what follows. Then, we study some particular classes of graphs that we can meet in philosophy and give some examples of them.

1.1. Graph theory: a brief history

The origin of graph drawings is not well known, but we have already found some of them in ancient China.

As in the terrestrial order, where the ancient Chinese used cryptography to restrict the transmission of certain messages to the initiates, communication with the heavenly forces required in China a form adapted to the addressee, and conversely the message of celestial forces could not take over the form of a terrestrial message. Thus came the *fu*, these graphs, most often Daoist, which allowed followers to enter into communication with spirits and spirits to communicate with the earthly world (see Figure 1.1).

In this region of the world, some graphic constructions may also be found in decorative motives of architecture or jails (see Figure 1.2) which follow certain logics (He and Schnabel, 2018).

1 For a color version of all the figures in this chapter, see http://www.iste.co.uk/parrochia/graphs.zip.

Figure 1.1. *Taoist Fu surmounted by a constellation of 5 stars (right); official of fate (left) (BNF, Coins, Medals and Antiques - CF A-135) (source: BNF)*

Figure 1.2. *Chinese decorative graphs (photo: Z. Guo)*

It seems that one of the earliest forms of them in western countries are probably that of Morris and Mill games, as shown in Figure 1.3, where the nodes of the graph drawing are the positions that game counters can occupy, the edges indicating how game counters can move between nodes (see Kruskal 1960, pp. 272–286).

Figure 1.3. *Depiction of Morris gameboards*

The origin of Mill gameboards comes from stone carvings in Ancient Egypt but the date of the earliest examples of Mill gameboards to be drawn in a book only goes back to the end of the 13th century.

However, early graph drawings were not exclusively an invention of Asia or the Middle East.

Quipus, which are bundles of cords whose knots and combinations replaced the writing (see Figure 1.4), were used by the Incas from the 13th to 16th centuries for accounting purposes and especially to register important facts and events. Of the few hundred surviving examples, roughly 25% exhibit a hierarchical structure and therefore qualify as tree drawings (see Ascher and Ascher 1997).

Figure 1.4. *Incas' quipu (source: Wikipedia article: Quipu)*

Another example of ancient graph drawings is family trees which decorated the atria of the patrician roman villas. Though no example has survived, it has been described by Pliny the Elder and Seneca. It was not until the Middle Ages that we saw them reappear.

During the 14th century, there appeared trees depicting the categories of different kinds (vices, virtues), while logicians often use them to represent abstract information (Porphyrian tree) or to illustrate the system of logical arguments (square of opposition).

After that, nothing comes before the famous paper of 1736 written by Leonhard Euler on the seven bridges of Kœnigsberg[1] and regarded as the first paper in the history of graph theory (see Biggs et al. 1986; Parrochia 1993b). This

1 The city of Kœnigsberg (East Prussia), today called Kaliningrad (Russia), being built around two islands located on the Pregel and connected by a bridge, six other bridges connect the banks of the river to each other. A famous problem was whether you could go around the city by crossing each bridge only once. L. Euler resolved this problem in the negative. This resulted in the notion of the "Eulerian cycle".

paper, as well as the one written by Vandermonde on the knight problem in chess, carried on with the dream of the so-called "analysis situs" initiated by Leibniz and the ancestor of topology.

Euler's formula relating the number of edges, vertices and faces of a convex polyhedron was studied and generalized by Cauchy (1813) and L'Huillier (1812-1813) and represents the beginning of this new branch of mathematics: algebraical topology.

More than one century after Euler's paper and while Listing was introducing the concept of topology, Cayley was led by an interest in particular analytical forms arising from differential calculus to study a particular class of graphs, the trees (see Cayley 1857). This study was partly motivated by important problems in theoretical chemistry. The techniques Cayley used mainly concerned the enumeration of graphs with specific properties. Enumerative graph theory then arose from his results and also the fundamental results published between 1935 and 1937. These were then generalized by De Bruijn in 1959. The links established by Cayley between his results on trees and contemporary studies of chemical composition (see Cayley 1875) influenced the development of a standard terminology in graph theory.

For example, the term "graph" was introduced by Sylvester in a paper published in 1878 in *Nature*, where he draws an analogy between "quantic invariants" and "co-variants" of algebra and molecular diagrams (see Sylvester 1878):

> Every invariant and co-variant thus becomes expressible by a graph precisely identical with a Kekuléan diagram or chemicograph. I give a rule for the geometrical multiplication of graphs, i.e. for constructing a graph to the product of in- or co-variants whose separate graphs are given.

The first textbook on graph theory was written by Dénes König and published in 1936 (see König 1990). Another book by Frank Harary, published in 1969, was "considered the world over to be the definitive textbook on the subject" (see Gardner 1992, p. 203) and enabled mathematicians, chemists, electrical engineers and social scientists to talk to each other. In France, Claude Berge (see Berge 1970) has also published in the 1970s a very important book about graphs and hypergraphs.

Let us now enter some particular problems in graph theory.

One of the most famous and stimulating problems in this field is the four color problem: "Is it true that any map drawn in the plane may have its regions colored with four colors, in such a way that any two regions having a common border have different colors?" This problem was first posed by Francis Guthrie in 1852 and its first written record is in a letter of De Morgan addressed to Hamilton the same year. Many incorrect proofs have been proposed, including those by Cayley, Kempe and others. The study and the generalization of this problem by Tait, Heawood, Ramsey and Hadwiger led to the study of the colorings of the graphs embedded on surfaces with arbitrary genus. Tait's reformulation generated a new class of problems, the factorization problems, particularly studied by Petersen and König. The works of Ramsey on colorations and more specially the results obtained by Turán in 1941 were at the origin of another branch of graph theory, extremal graph theory.

The four-color problem remained unsolved for more than a century. In 1969, Heinrich Heesch (see He and Schnabel 2018) published a method for solving the problem using computers. A computer-aided proof produced in 1976 by Kenneth Appel and Wolfgang Haken makes the fundamental use of the notion of "discharging" developed by Heesch (see Appel and Haken 1977a, 1977b). The proof involved checking the properties of 1,936 configurations by computer, and was not fully accepted at the time due to its complexity. A simpler proof considering only 633 configurations was given 20 years later by Robertson et al. (1997).

The autonomous development of topology from 1860 to 1930 fertilized graph theory back through the works of Jordan, Kuratowski and Whitney. Another important factor of common development of graph theory and topology came from the use of the techniques of modern algebra. The first example of such a use comes from the work of the physicist Gustav Kirchhoff, who published in 1845 his Kirchhoff's circuit laws for calculating the voltage and current in electric circuits.

The introduction of probabilistic methods in graph theory, especially in the study of Erdös and Rényi of the asymptotic probability of graph connectivity, gave rise to yet another branch, known as *random graph theory*, which has been a fruitful source of graph-theoretic results.

The consideration of flows and networks and the development of the Ford–Fulkerson method provide a new approach in graph theory, combinatorics and then algorithmics.

We will first explore the already large universe of graphs.

1.2. Basic definitions

Let us begin with some definitions.

DEFINITION 1.1.– *In its modern mathematical sense, a graph is an ordered pair $G(V, E)$ comprising a set V of vertices or nodes or points together with a set E of edges or arcs or lines, which are 2-element subsets of V (i.e. an edge is associated with two vertices, and that association takes the form of the unordered pair comprising those two vertices). This is not exactly, indeed, the most general definition. To avoid ambiguity, we will see later that this type of graph may be described precisely as undirected and simple. This characteristics are sufficient for the moment.*

REMARK 1.1.– *The vertices belonging to an edge are called the ends or end vertices of the edge. A vertex may exist in a graph and not belong to an edge.*

REMARK 1.2.– *The sets V and E are usually taken to be finite, and many of the well-known results are not true (or are rather different) for infinite graphs because many of the arguments fail in the infinite case.*

DEFINITION 1.2.– *The order of a graph is $|V|$, its number of vertices. The size of a graph is $|E|$, its number of edges. The degree or valency of a vertex is the number of edges that connect to it, where an edge that connects a vertex to itself (a loop) is counted.*

For an edge $\{x, y\}$, graph theorists usually use the somewhat shorter notation xy.

1.3. Different types of graphs

As we have seen, many graphs have been discovered throughout the history of the discipline, so that today, even restricting us to interesting or well-known graphs, which are already very numerous, we are dealing with an extremely large set, in fact a very bushy forest, not to say a real jungle.

Exploring this collection supposes that we have identified some distinctive features that can guide us and allow us to define classification criteria.

There are, in fact, various types of graphs depending upon the number of vertices, the number of edges, their interconnectivity and their overall structure. We will discuss only a certain few important types of graphs in this chapter.

DEFINITION 1.3 (NULL GRAPH).– *A graph having no edges is called a null graph.*

Philosophical example: At the end of his book *The Unique and His Property*, the hymn to egoism, the notion that the individual is the measure of all things, Max Stirner states "that the individual is a world history for himself, and possesses his property in the rest of the world's history". It ensures "that every higher essence above be it God, be it the human being, weakens the feeling of my uniqueness, and only before the sun of this awareness". On the contrary, he assures: "If I base my affair on myself, the unique, then it stands on the transient, the mortal creator, who consumes himself, and I may say: I have based my affair on nothing" (see Stirner 2017, pp. 376–377). We cannot better illustrate both the idea of a graph with a single vertex and the denomination of this one as "null graph".

DEFINITION 1.4 (TRIVIAL GRAPH).– *A graph with only one vertex is called a trivial graph.*

Philosophical example: The circular $J - set$, with one vertex and one loop, in Finsler's theory of sets (see Booth and Ziegler 1996), is a beautiful example of trivial graph[2].

DEFINITION 1.5 (NON-DIRECTED GRAPH).– *A graph which contains edges but in which the edges are not directed ones is a non-directed graph.*

DEFINITION 1.6 (DIRECTED GRAPHS).– *A graph in which each edge has a direction is a directed graph (sometimes, it may be called a network).*

DEFINITION 1.7 (SIMPLE GRAPH).– *A graph with no loops and no parallel edges is called a simple graph.*

DEFINITION 1.8 (REGULAR GRAPH).– *A graph G is said to be regular, if all its vertices have the same degree. In a graph, if the degree of each vertex is k, then the graph is called a k-regular graph.*

DEFINITION 1.9 (COMPLETE GRAPH).– *A simple graph with n mutual vertices is called a complete graph and it is denoted by K_n.*

REMARK 1.3.– *In other words, if a vertex is connected to all other vertices in a graph, then it is called a complete graph.*

2 On the metaphysical sense of the J-set, see Parrochia (1999).

DEFINITION 1.10 (CYCLE GRAPH).– *A simple graph with* n *vertices* ($n \geq 3$) *and* n *edges is called a cycle graph if all its edges form a cycle of length* n.

REMARK 1.4.– *If the degree of each vertex in the graph is two, then it is called a cycle graph.*

Philosophical example: As in the system of Hegel, the sequence of the main concepts of the philosophy of Eric Weil forms a cycle graph.

DEFINITION 1.11 (WHEEL GRAPH).– *A wheel graph is obtained from a cycle graph* C_{n-1} *by adding a new vertex. That new vertex is called a Hub which is connected to all the vertices of* C_n.

Philosophical example: In Leibnizian philosophy, God is a *Hub* because He is connected to all the monads. Conversely, each monad is connected to all aspects of the world. So, the Leibnizian system is, from a graphic point of view, a wheel of wheels.

DEFINITION 1.12 (CYCLIC GRAPH).– *A graph with at least one cycle is called a cyclic graph.*

Philosophical example: There are a lot of cycles in the classic philosophical systems of Leibniz or Hegel.

DEFINITION 1.13 (ACYCLIC GRAPH).– *A graph with no cycles is called an acyclic graph.*

Philosophical example: In Descartes' philosophy, the "long chains of reasons" are examples of acyclic graphs. But Descartes' system itself is not acyclic (see Beyssade 1979; Parrochia 1993b, p. 183, 2012, pp. 114–115).

DEFINITION 1.14 (BIPARTITE GRAPH).– *A simple graph* $G = (V, E)$ *with a vertex partition* $V = \{V_1, V_2\}$ *is called a bipartite graph if every edge of* E *joins a vertex in* V_1 *to a vertex in* V_2.

Philosophical example: According to Gilles Deleuze (see Deleuze 2003, pp. 42–43), in the whole graph of Spinoza's *Ethics*, we can make a partition into two big classes of vertices associated with statements: on the one hand, the propositions, corollaries and proofs, which form the deductive part of the book; on the other hand, the scholia, a discontinuous broken chain which do not belong to the deduction. However, scholia are connected to propositions and corollaries (or their proofs) by some edges. All that do not constitute really a bipartite graph.

DEFINITION 1.15 (COMPLETE BIPARTITE GRAPH).– *A bipartite graph $G = (V, E)$ with partition $V = \{V_1, V_2\}$ is said to be a complete bipartite graph if every vertex in V_1 is connected to every vertex of V_2.*

DEFINITION 1.16 (STAR GRAPH).– *A complete bipartite graph of the form $K_{1,n-1}$ is a star graph with n vertices. A star graph is a complete bipartite graph if a single vertex belongs to one set and all the remaining vertices belong to the other set.*

1.4. More on the list of graphs

As far as we know, there is no real classification of graphs. However, we can present a list of well-known graphs that we can put together according to their name:

1) First, a catchy "name" for a graph often comes from some drawing of the graph. A first part of the "graph menagerie" is formed by the most simple graphs. For example, among the graphs with at most five vertices, we get the following: triangle, claw, paw, kite, house, bull, bowtie, dart, etc. These simple graphs are sometimes used to decompose more complex graphs or to identify graphs which do not include them: for example, triangle-free graphs, paw-free graphs, etc.

2) The discovery of some graphs happened in the development of graph theory in connection with some particular problems (e.g. a counterexample to a conjecture or the solution of some extremal problem). In this context, we meet some famous graphs named after the mathematician who found them.

Here are, for example, the following: Petersen graph, Balaban 10-cage, Dürer graph, Herschel graph, Sousselier graph, Sylvester graph, Tutte graph, Wagner graph, etc. No need to define them for now: we will do so when we encounter them if this is the case.

3) We can also identify some symmetric or regular graphs: highly symmetric graphs, strongly regular graphs, symmetric graphs, semi-symmetric graphs, etc. A graph which is not symmetric is an asymmetric graph. We will be concerned with some of them.

4) There are also big graph families, with well-characterized properties: complete graphs, complete bipartite graphs, cycles, friendship graphs, fullerene graphs, platonic solids, truncated solids, snarks, stars, wheel graphs, etc.

Some websites like "The house of graphs" give a catalog of interesting graphs with particular properties: for example, a certain number of vertices, edges, cycles or some more sophisticated properties like the diameter, the girth, the chromatic number or the number of cliques, automorphisms, etc.

When you look for a particular graph, you can guess it has always existed and that its properties are already well known. If not, you have a chance to be the discoverer of a new class of graphs.

An important point is that a graph generally has different kinds of drawings, but all of these are isomorphic. So, a graph is in fact a class of graphs generally defined up to an isomorphism.

So far, philosophers, for structuring thought, have used only a very small part of the possibilities listed in mathematics, especially in graph theory.

Generally, a few concepts (often less than 20) are enough for a philosopher to allow them to represent the world. So far, philosophers have only explored one possibility each time – in other words, the choice was made blind.

But today, we can, by combinatorial enumeration, generate all possible graphs containing a given number of vertices and edges. The results defy the imagination. Thus, for a number of vertices less than 20, the number of graphs likely to be generated is given below (this sequence S corresponds to graphs with 0 vertices, graphs with 1 vertex, graphs with 2 vertices, etc.). The last number is the number of graphs with 19 vertices:

$$S = 1, 1, 2, 4, 11, 34, 156, 1044, 12346, 274668, 12005168, 1018997864,$$
$$165091172592, 50502031367952, 29054155657235488,$$
$$31426485969804308768, 64001015704527557894928,$$
$$245935864153532932683719776, 1787577725145611700547878190848,$$
$$24637809253125004524383007491432768.$$

To give a speaking example of what we are putting forward, here is a small sample of the graphs $G(X, U)(|X| = 6)$ that we can construct. Many different architectonics could thus be generated from the same basic corpus (see Figure 1.5).

A philosopher may say that this list has no more interest than the *One Hundred Billion Poems* by Queneau (1961) and could well be, like the late research of Oulipo, a free game where content gets lost. What obviously makes a good philosopher, like what makes a good poet, probably has little to do with combinatorics. But the counts remain interesting and put our constructions into perspective. They should, at least, make those who take themselves for masters more modest.

So, as we can see, there exist an immense zoo of graphs, where we can find, indeed, a lot of non-trivial forms of conceptual architectonic structures.

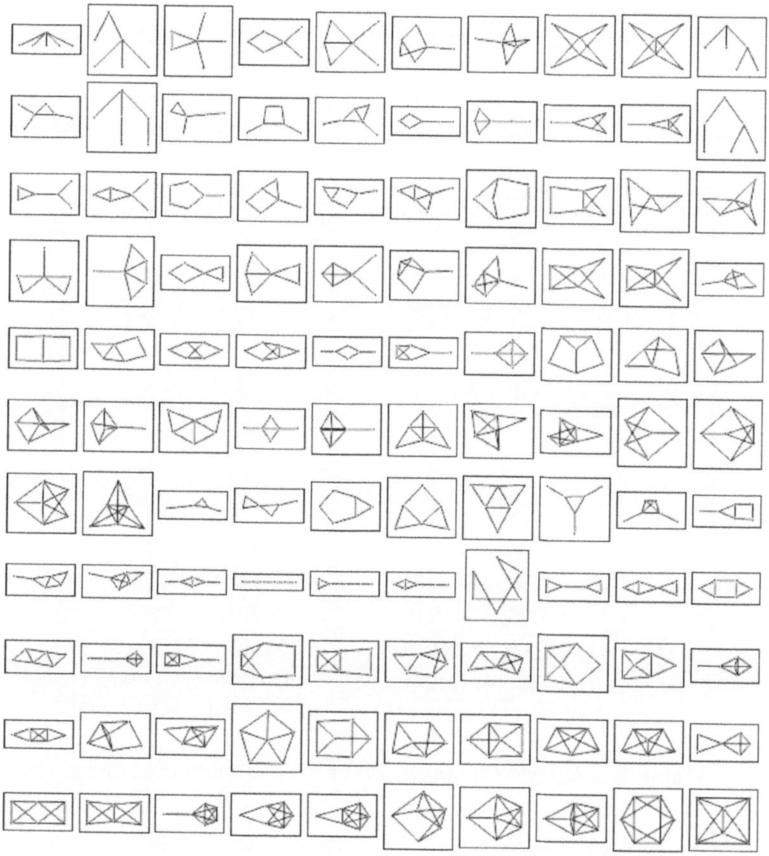

Figure 1.5. *Some graphs* $G(X, U)$ *with* $|X| = 6$

1.5. Graphs and vertices

Once we know whether the graph is labeled or not, the simplest criteria that can be retained are obviously the number of vertices. The number of the first graphs of unlabeled vertices is given in Table 1.1. Let v be the number of vertices of a graph, n be the number of connected graphs and m be the number of all possible graphs. As we can see, for $v = 6, n, m > 100$ and for $v > 7$, these numbers are growing extremely fast. For $v = 16$, these very numbers challenge the imagination. But it is not uncommon to deal with graphs whose number of vertices are greater. For a philosopher, for example, who has to organize the relations between 20 categories or more, this is quite possible. This tends to call into question the possibility of constructing a philosophy on a deductive mode: namely, from a finite set of speculative categories, attempting to try every possible arrangement to choose the best one. Of course, we cannot proceed like this.

Indeed, we think it could be useful to a philosopher, to know a few things about some simple graphs that could be bases of some cognitive systematization.

Vertices	Connected graphs	All graphs
1	1	1
2	1	2
3	2	4
4	6	11
5	21	34
6	112	156
7	853	1044
8	11117	12346
9	261080	274668
10	11716571	12005168
11	1006700565	1018997864
12	164059830476	165091172592
13	50335907869219	50502031367952
14	29003487462848061	29054155657235488
15	31397381142761241960	31426485969804308768
16	63969560113225176176277	64001015704527557894928

Table 1.1. *Number of graphs with a fixed number of vertices*

The main graphs we have to deal with in philosophy are connected graphs, though some philosophical doctrines, based on a collection of aphorisms, present a disconnected structure. Note also that, since Nietzsche and Kierkegaard, most of the philosophers have abandoned (rightly or wrongly – maybe wrongly –) the project to construct a *system* in the classical sense, i.e. a coherent and definitive organization of the whole of thought based on principles of economy and harmony (for details, see section 2.5 of Chapter 2, and also Chapter 5).

DEFINITION 1.17.– *A graph is said to be connected in the sense of a topological space if there is a path from any point to any other point in the graph. A graph that is not connected is said to be disconnected.*

REMARK 1.5.– *This definition means that the null graph and the singleton graph are considered connected, while empty graphs on $n \geq 2$ nodes are disconnected.*

All the connected graphs $G(V, E)$ with $|V| \leq 5$ are shown in Figure 1.6.

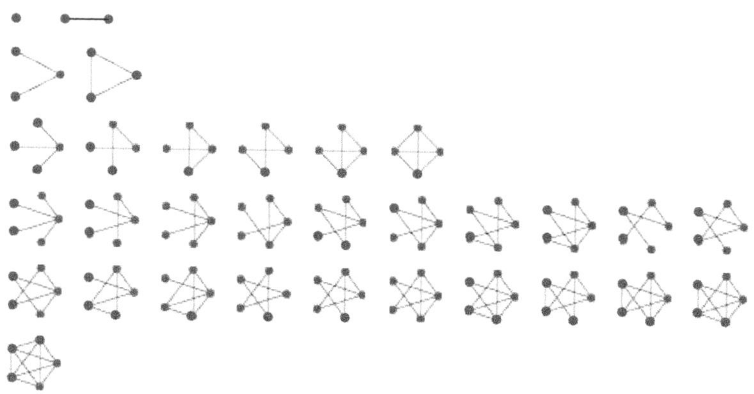

Figure 1.6. *Connected graphs $G(V, E)$ with $V \leq 5$*

1.6. Some operations on graphs

DEFINITION 1.18.– *The complement or inverse of a graph G is a graph H on the same vertices such that two distinct vertices of H are adjacent if and only if they are not adjacent in G.H is sometimes written as $\bar{G}$.*

THEOREM 1.1.– *If G is disconnected, then its complement $\bar{G}$ is connected.*

REMARK 1.6.– *However, the converse is not true, as can be seen using the examples of the 4-vertex path graph and the 5-vertex cycle graph which are both connected and isomorphic to their complement.*

Here are some binary operations between two simple graphs $G_1 = (V_1, E_1)$ and $G_2 = (V_2, E_2)$:

1) The union is $G_1 \cup G_2 = (V_1 \cup V_2, E_1 \cup E_2)$ (simple graph).

2) The intersection $G_1 \cap G_2 = (V_1 \cap V_2, E_1 \cap E_2)$ (simple graph).

3) The ring sum $G_1 \oplus G_2$ is the subgraph of $G_1 \cup G_2$ induced by the edge set $E_1 \oplus E_2$ (simple graph).

The examples are presented in Figure 1.7.

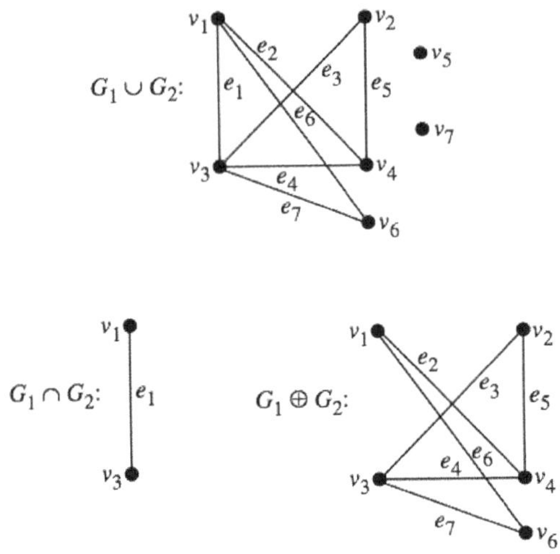

Figure 1.7. *Three binary operations between two simple graphs*

NOTE.– The set operation $\oplus$ is the symmetric difference, i.e.

$$E_1 \oplus E_2 = (E_1 - E_2) \cup (E_2 - E_1).$$

Since the ring sum is a subgraph induced by an edge set, there are no isolated vertices. All three operations are commutative and associative.

The operations $\cup$, $\cap$ and $\oplus$ can also be defined for more general graphs other than simple graphs. Naturally, we have to "keep track" of the multiplicity of the edges. In this case:

1) $\cup$: The multiplicity of an edge in $G_1 \cup G_2$ is the larger of its multiplicities in G_1 and G_2.

2) $\cap$: The multiplicity of an edge in $G_1 \cap G_2$ is the smaller of its multiplicities in G_1 and G_2.

3) $\oplus$: The multiplicity of an edge in $G_1 \oplus G_2$ is $|m_1 - m_2|$, where m_1 is its multiplicity in G_1 and m_2 is its multiplicity in G_2.

1.7. Graph isomorphisms

1.7.1. *Self-complementary graphs*

DEFINITION 1.19.– *A graph which is isomorphic to its graph complement is called a self-complementary graph (see Figure 1.8).*

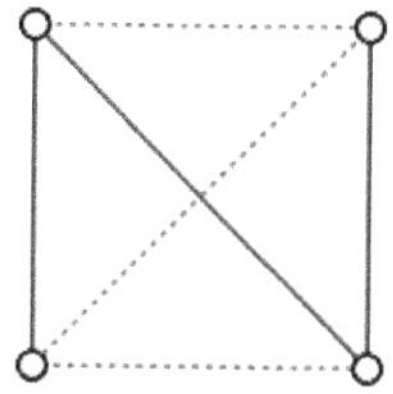

Figure 1.8. *An example of a self-complementary graph*

The numbers of simple self-complementary graphs on n = 1, 2, ... nodes are 1, 0, 0, 1, 2, 0, 0, 10, ... The first of them is shown in Figure 1.9.

Every self-complementary graph is not only connected, but also traceable. A trivial example of a self-complementary graph in philosophy is the world's image of the line in Plato's *Republic* (510e–511a), which can be associated with P_4, while the self-complementary graph C_5 is the figure circumscribing the Pythagorean pentagram.

Here, we allow ourselves to draw the attention of philosophers to an important methodological point.

The existence of self-complementary graphs proves that it is not always relevant to take the complement of a conceptual graph, contrary to the method advocated by Jacques Derrida and implemented in his work on the notion of "parergon" (see Derrida and Owens 1979)[3]. If the graph in question is a self-complementary graph, then, it is absolutely the same, from the structural point of view, to study the conceptual graph or its complementary.

3 The English version is a translation of section II of the four-part essay entitled "Parergon" published in Derrida 1978. Parts of this text originally appeared in the French review *Digraphe* 2 (see Derrida 1974 and our commentary in Parrochia 1993a). In Derrida's idea, what is at the edge of a philosopher's work and seems to fall outside their official theses sometimes allows them to be reinterpreted to the point of revealing what their discourse tends to veil. In the case of self-complementary graphs, it is not clear whether the so-called "parergonal border logic" could possibly reveal more than the graph itself.

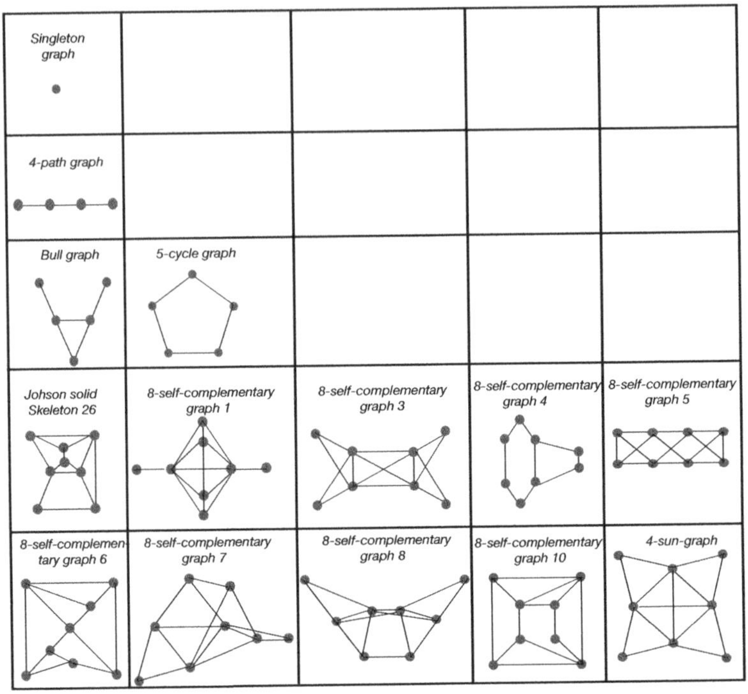

Figure 1.9. *The first self-complementary graphs*

1.7.2. *Properties of self-complementary graphs*

For a self-complementary graph on $n \geq 5$ vertices, there is a graph cycle of length ℓ for every integer $3 \leq \ell \leq n - 2$ (see Rao 1977; Farrugia 1999, p. 51). As a result, the graph circumference of a self-complementary graph on $n > 5$ vertices is either n (i.e. the graph is Hamiltonian), $n - 1$ or $n - 2$.

By definition, a self-complementary graph must have exactly half the total possible number of edges, i.e. $n(n-1)/4$ edges for a self-complementary graph on n vertices. Since $n(n-1)$ must be divisible by 4, it follows that $n = 0$ or $1(\text{mod.}4)$.

We may also remark that there is a polynomial-time condition to determine if a self-complementary graph is Hamiltonian.

And as the number of self-complementary graphs on n nodes can be derived from Pólya's enumeration theorem, using self-complementary graphs in philosophy should allow us to process a kind of deductive philosophy.

1.8. Symmetric and asymmetric graphs

Once we have seen whether a graph is finite or infinite, whether it is directed or not, simple or not, etc., we may search whether a graph is symmetric or not. The symmetry property is characteristic of mathematical graphs and leads to beautiful figures. But there is no reason for a philosophical graph, i.e. a graph which represents the whole world or some part of it, to be symmetric[4]. Moreover, symmetry is not a basic property. It is a consequence of more fundamental properties like the fact of being edge-transitive and vertex-transitive. Let us explain that with the help of some definitions.

DEFINITION 1.20.– *An edge-transitive graph is a graph G such that, given any two edges* e_1 *and* e_2 *of G, there is an automorphism of G that maps* e_1 *to* e_2.

DEFINITION 1.21.– *A vertex-transitive graph is a graph G in which, given any two vertices* v_1 *and* v_2 *of G, there is some automorphism:*

$$f : V(G) \to V(G)$$

such that:

$$f(v_1) = v_2$$

DEFINITION 1.22 (SYMMETRIC GRAPH).– *A symmetric graph is a graph that is both edge- and vertex-transitive (Holton and Sheehan 1993, p. 209). However, care must be taken with this definition since arc-transitive or 1-arc-transitive graphs are sometimes also known as symmetric graphs (Godsil and Royle 2001, p. 59).*

Of course, edge-transitive graphs are not necessarily vertex-transitive. Some graphs like the Folkman graph (20 vertices) (see Figure 1.10), the Gray graph (54 vertices), the Iofinova–Ivanov graph (110 vertices), the Ljubljana graph (112 vertices) and the Tutte-12 cage (126 vertices) are examples of such a case.

Conversely, there exist vertex-transitive graphs which are not edge-transitive. The graph corresponding to a triangular prism (see Figure 1.11), for example, is

4 We know that natural evolution seems to show a preference for symmetric structures, which can be explained for algorithmic reasons: they are easier to obtain at random than asymmetric structures, which are generally much more complex (see Louis et al. 2022). But in the evolution of thought, and so, in philosophy, no such algorithmic process exists and "simple" forms do not seem to be favored. However, for reasons of comprehension as much as of aesthetics, the system must surely be presented in a form that is economical and quick to use.

vertex-transitive but not edge-transitive. More generally, two copies G, G' of the complete graph K_n for $n \geq 3$ that are linked by n edges (one edge between v_i and v'_i for each i) is vertex-transitive but not edge-transitive.

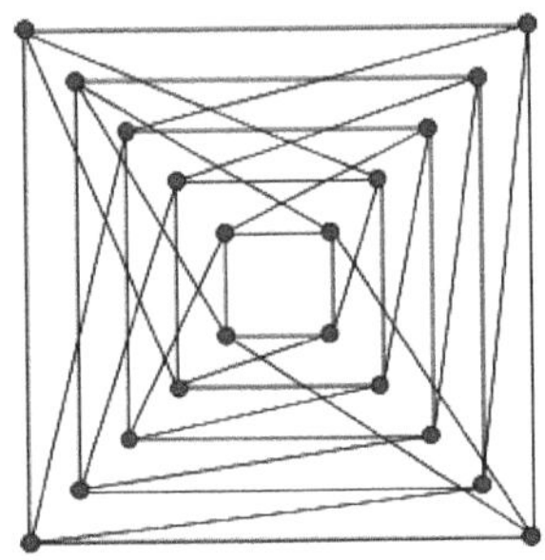

Figure 1.10. *The Folkman graph (source: Wikipedia article: Folkman graph (modified from Wolfram Mathworld))*

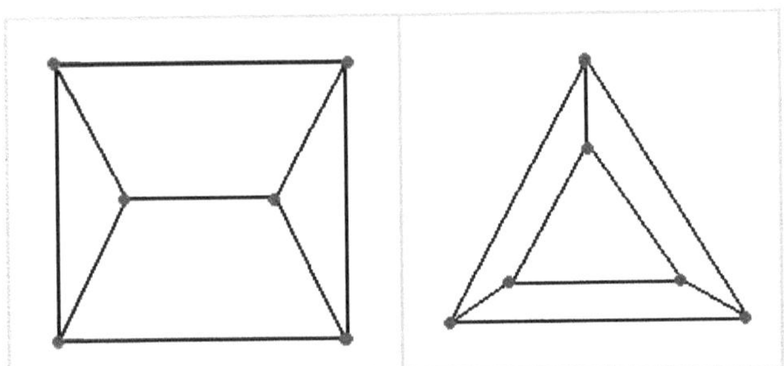

Figure 1.11. *The triangular prism*

DEFINITION 1.23 (ASYMMETRIC GRAPH).– *A graph which is not symmetric is said to be asymmetric. An asymmetric graph has no nontrivial symmetries, i.e. it is a graph for which there are no other automorphisms than the trivial automorphism (the identity mapping of the graph onto itself).*

An example is the Frucht graph (12 vertices, 18 edges), a 3-regular pancyclic Halin graph first described by Robert Frucht in 1939 (see Figure 1.12).

Can we measure the degree of asymmetry of a graph? The answer is positive. As Erdös and Réniy have shown (see Erdös and Rényi 1963), any asymmetric graph can be made symmetric by deleting some of its edges and by adding certain new edges connecting its vertices. This is called a *symmetrization* of G. Now, let us take the sum of the number of deleted edges – say r – and the number of new edges – say s. The degree of asymmetry $A[G]$ of a graph G is the minimum of $r + s$ where the minimum is taken over all possible symmetrizations of the graph G. If in order to make a graph symmetric, we delete r of its edges and add s new edges, we shall say that we changed $r + s$ edges. Clearly, the asymmetry of a symmetric graph is according to this definition equal to 0, while the asymmetry of any asymmetric graph is a positive integer. The question arises: how large can the degree of asymmetry of a graph of the order n (i.e. a graph which has n vertices) be? We shall denote by $A(n)$ the maximum of $A[G]$ for all graphs G of order n ($n = 2, 3, ...$). We put further $A(1) = +\infty$. The authors prove that for $n = 2$ to 5, $A(n) = 0$. Some graphs can be made symmetric by adding one edge (Figure 1.13(a)) or by omitting one edge (Figure 1.13(b)).

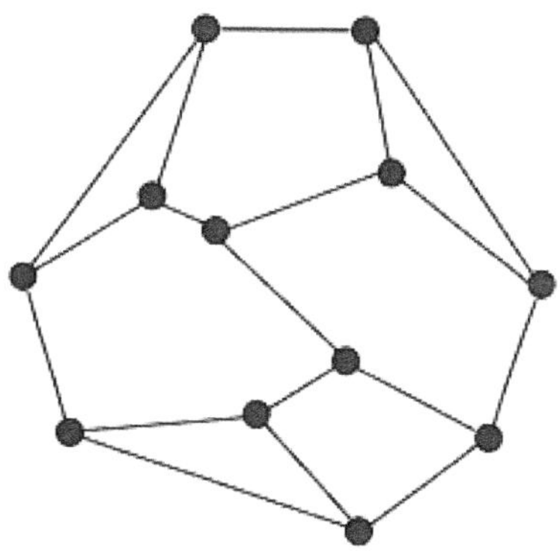

Figure 1.12. *The Frucht graph*

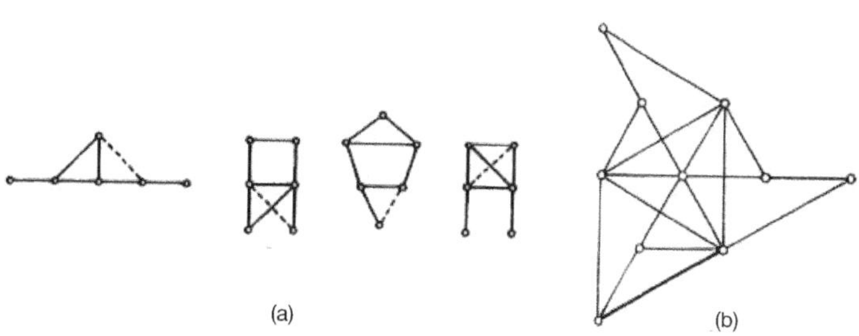

Figure 1.13. *Symmetrization of some graphs (from Erdös and Renyi 1963, p. 297).*

The authors also get the following results:

THEOREM 1.2.– *The asymmetry of a graph of order n cannot exceed $\frac{n-1}{2}$ if n is odd, while if n is even the asymmetry of a graph of order n cannot exceed $\frac{n}{2} - 1$.*

This is the best estimation, so we have, in other words:

THEOREM 1.3.–

$$\lim_{n\to\infty} \frac{A(n)}{n} = \frac{1}{2}.$$

The degree of asymmetry of a graph is therefore at most slightly lower than the number of its edges. In other words, if the world were an asymmetric graph of order n, it would be necessary at most to change a little less than half of the number of its edges to transform it into a symmetric graph.

THEOREM 1.4.– *The probability that a graph with n vertices and N edges chosen at random should be asymmetric tends to be 1 for $n \to \infty$.*

THEOREM 1.5.– *Assume now that Γ is an infinite graph. The probability P that Γ should be symmetric is such that $P = 1$.*

These last theorems mean that almost all finite graphs are asymmetric, while almost all infinite graphs are symmetric. So, if the world graph is finite, it has a big chance of being asymmetric, while if it is infinite, it has a big chance of being symmetric.

Perhaps there exist some exceptions. In Spinoza's philosophy, the infinite modes of thought and extension (and the unknown others) are infinite chains, so they are infinite edge-transitive graphs. But the whole graph collecting these chains is not connected. And so, it is not vertex-transitive. However, there is a correspondence between any pairs of chains (*Ethic*, book 2, prop. 7), which is just a consequence of the fact that such a graph of a bichain is necessarily bipartite (see Biggs 1993, p. 118).

1.9. Extremal graphs

In this part of mathematics called "analysis", we know that it can be very useful to study the extrema of a function. We draw from it information about its derivative and the direction of variation of the function. Extrema are critical points which allow us to characterize it.

In the same way, this part of graph theory named *extremal graph theory* studies extremal (maximal or minimal) graphs that satisfy a certain property, which allow us to get a better understanding of the graph. Extremality can be taken with respect to different graph invariants, such as order, size or girth. More abstractly, it also studies how global properties of a graph influence the local substructures of it.

For example, a question like "which acyclic graphs on n vertices have the maximum number of edges?" is typically an extremal graph theory question. The extremal graphs which answer this question are trees on n vertices with $n - 1$ edges. More generally, questions are of the following type: given a graph property P, an invariant u and a set of graphs S, we wish to find the minimum value of m such that every graph in S with $u > m$ possesses property P. In the example above, S was the set of n-vertex graphs, P was the property of being cyclic and u was the number of edges in the graph. Thus, every graph on n vertices with more than $n - 1$ edges must contain a cycle.

Several foundational results in extremal graph theory are questions of the above-mentioned form. For instance, the question of the largest number of edges that we can add to an n-vertex graph without creating clique of size k is answered by Turán's theorem. Instead of cliques, if the same question is asked for complete multi-partite graphs, the answer is given by the Erdös–Stone theorem.

Historically, extremal graph theory, in its strictest sense, has been developed by Hungarians. It started in 1941 when Turán proved his theorem determining those graphs of order n, not containing the complete graph K_k of order k, and extremal with respect to size (i.e. with as many edges as possible).

Turán's theorem answers the following question: what is the maximum possible number of edges in an undirected graph G with n vertices which does not contain K_3 (the graph with three vertices A, B, C and three edges AB, AC, BC, i.e. a triangle) as a subgraph? The complete bipartite graph where the partite sets differ in their size by at most 1 is the only extremal graph with this property. It contains:

$$\left\lfloor \frac{n^2}{4} \right\rfloor$$

edges.

Similar questions have been studied with various other subgraphs H instead of K_3. For instance, the Zarankiewicz problem concerns the largest graph that does not contain a fixed complete bipartite graph as a subgraph, and the even circuit theorem concerns the largest graph without a fixed-length even cycle. Turán also found the (unique) largest graph not containing K_k which is named after him, namely the Turán graph (see Figure 1.14). This graph is the complete join of $k-1$ independent sets (as equisized as possible) and has at most:

$$\left\lfloor \frac{(k-2)n^2}{2(k-1)} \right\rfloor = \left\lfloor \left(1 - \frac{1}{k-1}\right) \frac{n^2}{2} \right\rfloor$$

edges. For C_4 (the square), the largest graph on n vertices not containing C_4 has:

$$\left(\frac{1}{2} + o(1)\right) n^{3/2}$$

edges.

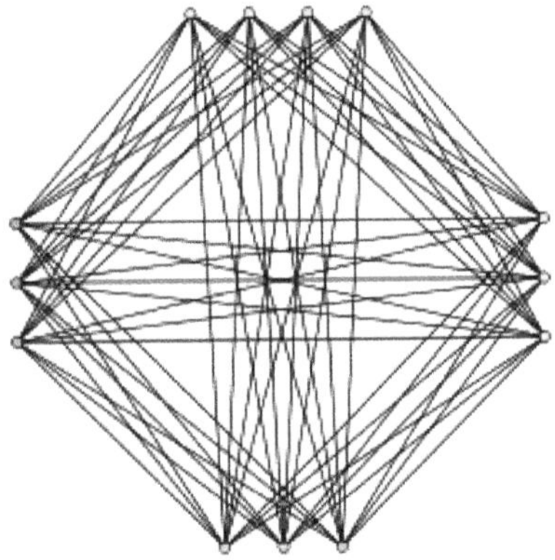

Figure 1.14. *Turán's graph (source: Wikipedia article: Turan graph)*

The preceding theorems give conditions for a small object to appear within (perhaps) a very large graph. At the opposite extreme, we might search for conditions which force the existence of a structure which covers every vertex. But it is possible for a graph with:

$$\binom{n-1}{2}$$

edges to have an isolated vertex – even though almost every possible edge is present in the graph – which means that even a graph with a very high density may have no interesting structure covering every vertex. Simple edge counting conditions, which give no indication as to how the edges in the graph are distributed, thus often tend to give uninteresting results for very large structures.

Instead, we introduce the concept of *minimum degree*. The minimum degree of a graph G is defined to be:

$$\delta(G) = \min_{v \in G} d(v).$$

Specifying a large minimum degree removes the objection that there may be a few "pathological" vertices. If the minimum degree of a graph G is 1, for example, then there can be no isolated vertices (even though G may have very few edges).

A classic result in this context is Dirac's theorem, which states that every graph G with n vertices and minimum degree at least $n/2$ contains a Hamilton cycle.

This last result contains very interesting information for philosophers who want to construct philosophical systems with (or without) the Hamilton cycle.

1.10. Independence, non-separability, reconstruction conjecture

Regarding recent philosophy, there is one observation that the mathematician must make: much of what philosophers believe is generally inaccurate or erroneous. Indeed, most of the statements that the philosophical current known as "French theory" have been able to express are either trivial or false.

For example, the notion of "difference" and the privilege of simulacra, in Deleuze, is a kind of poor Heraclitism which seems to ignore the scientific notion of the invariant. But an antireflexive, antisymmetric and antitransitive world is hardly plausible: any theory would be stillborn there[5]. We can also object that there are scientific laws, and that the question is not that everything changes all the time in the world (trivial observation). It is rather to detect what does not change in what is changing, which is precisely the task of reason.

Likewise, one of Jacques Derrida's major ideas, which consisted of introducing the concept of "deconstruction", has no real relevance. Initial translation of the German words "Destruktion" and "Abbau" used by Heidegger in "Being and Time", and essentially intended to reintegrate time into the philosophical problematic, this concept took on a whole new meaning with

5 It will be observed that: 1) the simulacrum constantly differs from itself (see Plato's "sophist", or even Diderot's "nephew of Rameau" (see Parrochia 2006, pp. 167–173)); 2) as it does not resemble itself, nor can it resemble another (it is therefore also defined by antisymmetry). Resemblance not being, in any case, transitive, the simulacrum cannot be integrated into any transitive relation whatsoever. It could not therefore exist – if it had the least possible existence – only in an antireflexive, antisymmetrical and antitransitive universe.

Derrida's *deconstruction*. The philosopher has in fact used this word to signify an operation aimed at revealing the structure or the traditional architecture of the founding concepts of the whole of ontology or of Western metaphysics. The critique encompasses not only the analyzed object, the analyst and their language, but is supposed to show an inextricable entanglement of the content and the container of the analytical production in which everything in itself, reduced to a pure system of signs, finally reveals its nothingness in the frameworks drawn by classical metaphysics. It is therefore the whole of philosophy that collapses to make room for a commentary which, indefinitely, will exhaust itself in saying this impossibility[6].

Of course, mathematics – and in particular graph theory – is capable of finely formalizing the subtleties likely to be encountered in certain real situations. Thus, the notions of dependence and independence, of separability and inseparability, of composition and decomposition then take on a precise meaning and can no longer be confused. The philosophical world of "the more or less" can only give way, in the long term, to the scientific universe of precision.

DEFINITION 1.24 (INDEPENDENT SET).– *An independent set (also named "stable set", "coclique" or "anticlique") is a set of vertices in a graph, no two of which are adjacent. That is, it is a set I of vertices such that for every two vertices in I, there is no edge connecting the two. The size of an independent set is the number of vertices it contains.*

A *maximal independent set* is an independent set that is not a proper subset of any other independent set. The problem of finding such a set is called "the maximum independent set problem" and is an NP-hard optimization problem. As such, it is unlikely that there exists an efficient algorithm for finding a maximum independent set of a graph. Figure 1.15 gives an example of a maximal independent set in the Petersen graph.

If we see the whole of ontology or of Western metaphysics as a very complicated graph, we can guess it contains some independent sets, and conjecture that there is, among them, a maximal one. This possible existence of stability seems to have escaped the supporters of the "French theory", too eager to put an end to metaphysics.

Let us take a glance now to separable and unseparable graphs.

6 Michel Foucault immediately saw this inevitable consequence of Derrida's theses (see Foucault 1972).

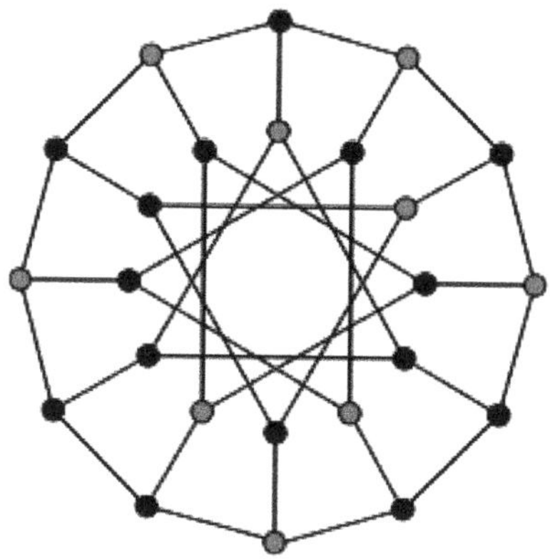

Figure 1.15. *Petersen graph GP(12, 4) (source: Wikipedia article: Independent set (graph theory): Generalized Petersen graph GP(12,4))*

DEFINITION 1.25 (CUT).– *A vertex v of a graph G is a cut vertex or an articulation vertex of G if the graph $G - v$ consists of a greater number of components than G (see Figure 1.16).*

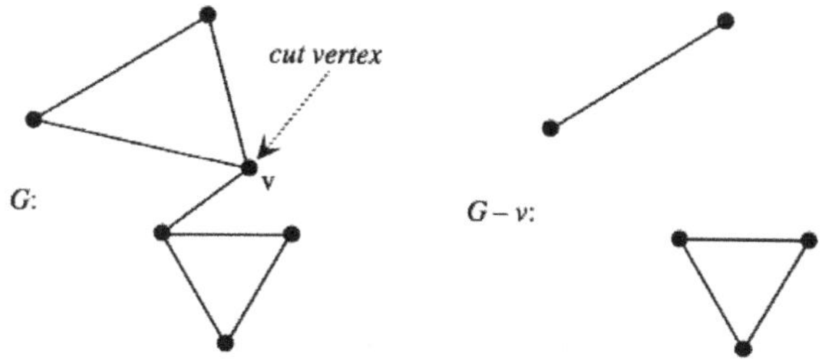

Figure 1.16. *A separable graph*

DEFINITION 1.26.– *More explicitly, we can say that a graph is separable if it is not connected or if there exists at least one cut vertex in the graph. Otherwise, the graph is non-separable (see Figure 1.17).*

Even when graphs are non-separable, they always admit subgraphs. Informally speaking, the reconstruction conjecture says that graphs are determined uniquely by their subgraphs. Let us introduce some new definitions to be able to express this more rigorously.

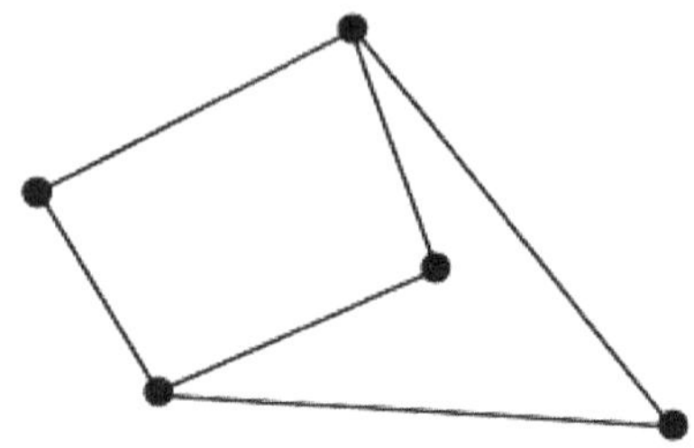

Figure 1.17. *A non-separable graph*

DEFINITION 1.27 (VERTEX-DELETED SUBGRAPH).– *Given a graph $G = (V, E)$, a vertex-deleted subgraph of G is a subgraph formed by deleting exactly one vertex from G. By definition, it is an induced subgraph of G.*

DEFINITION 1.28 (VERTEX-DECK).– *For a graph G, the vertex-deck (or simply "deck") of G, denoted $D(G)$, is the multiset of isomorphism classes of all vertex-deleted subgraphs of G. Each graph in $D(G)$ is called a card.*

DEFINITION 1.29 (HYPOMORPHIC GRAPHS).– *Two graphs that have the same deck are said to be hypomorphic.*

With these definitions, the conjecture can be stated like that.

Reconstruction conjecture: Any two hypomorphic graphs with at least three vertices are isomorphic.

REMARK 1.7.– *The requirement that the graphs have at least three vertices is necessary because both graphs with two vertices have the same decks.*

Harary (1964) suggested a stronger version of the conjecture.

Set reconstruction conjecture: Any two graphs with at least four vertices with the same sets of vertex-deleted subgraphs are isomorphic.

DEFINITION 1.30 (EDGE-DELETED SUBGRAPH).– *Given a graph $G = (V, E)$, an edge-deleted subgraph of G is a subgraph formed by deleting exactly one edge from G.*

DEFINITION 1.31 (EDGE-DECK).– *For a graph G, the edge-deck of G, denoted $ED(G)$, is the multiset of all isomorphism classes of edge-deleted subgraphs of G. Each graph in $ED(G)$ is called an edge-card.*

Edge reconstruction conjecture (see Harary 1964): any two graphs with at least four edges and having the same edge-decks are isomorphic.

As we can see, the "reconstruction conjecture" can be seen as a kind of methodological opposite to the famous Derrida's "deconstruction conjecture", a theory which, for lack of a rational method using effective formal tools, does not allow us to get any basic elements of the onto-theological discourse and to manage any new discovery within the frame of metaphysical knowledge. At the same time, it does not allow us to leave traditional metaphysics either and can only carry it out indefinitely, by limiting itself, here and there, to pointing out a gap between the texts and their contemporary criticism that nothing, ever, does not just fill.

On the contrary, the "reconstruction conjecture" (for a survey or the problem, see Harary 1974; Nash-Williams 1978) reveals in the graphs of metaphysics the subgraphs which generate them and, suddenly, allow, once combined differently, to escape the old determinations.

We hope that this method can be implemented in philosophy and will find full use of its immense possibilities.

This totally new approach to philosophy can be further explored by noting the following.

DEFINITION 1.32 (RECOGNIZABLE PROPERTIES).– *In the context of the reconstruction conjecture, a graph property is called recognizable if we can determine the property from the deck of a graph.*

The following properties of graphs are recognizable: order of a graph, number of edges, degree sequence of vertices, vertex-connectivity, Tutte polynomial, planarity, types of spanning trees and chromatic polynomial.

Both the reconstruction and set reconstruction conjectures have been verified for all graphs with at most 11 vertices by Brendan McKay (1997).

In a probabilistic sense, it has been shown by Béla Bollobás that almost all graphs are reconstructible (see Bollobás 1990). This means that the probability that a randomly chosen graph with n vertices is not reconstructible goes to 0 as n goes to infinity. In fact, it was shown that not only are almost all graphs reconstructible, but in fact that the entire deck is not necessary to reconstruct them: almost all graphs have the property that there exist three cards in their deck that uniquely determine the graph.

The conjecture has been verified for a number of infinite classes of graphs (and, trivially, their complements): regular graphs, trees, disconnected graphs, unit interval graphs, separable graphs without end vertices, maximal planar graphs, maximal outerplanar graphs, outerplanar graphs and critical blocks.

We get the following theorems:

THEOREM 1.6.– *The reconstruction conjecture is true if all 2-connected graphs are reconstructible (see Yang 1988).*

THEOREM 1.7 (DUALITY).– *If G can be reconstructed from its vertex deck $D(G)$, then its complement G' can be reconstructed from $D(G')$ as follows: Start with $D(G')$, take the complement of every card in it to get $D(G)$, use this to reconstruct G, then take the complement again to get G'.*

REMARK 1.8.– *Edge reconstruction does not obey any such duality: indeed, for some classes of edge-reconstructible graphs, it is not known if their complements are edge reconstructible.*

REMARK 1.9.– *Some structures are not generally reconstructible. Such are digraphs*[7]*, hypergraphs (see Kocay 1987), infinite graphs*[8]*, locally finite graphs*[9]*.*

Despite these limitations, the concept of "reconstruction" and the approach that goes with it are interesting. By taking them as far as they can lead in philosophy, we would perhaps be able to think from new structures, not only non-logocentric, but also non-Derridian ones, thus leaving definitively the cesspool where Derrida wants to lock us[10].

---

7 Infinite families of non-reconstructible digraphs are known, including tournaments (see Stockmeyer 1977) and non-tournaments (see Stockmeyer 1981). A tournament is reconstructible if it is not strongly connected (see Harary and Palmer 1967). A weaker version of the reconstruction conjecture has been conjectured for digraphs.

8 Let T be a tree with an infinite number of vertices such that every vertex has infinite degree, and let nT be the disjoint union of n copies of T. These graphs are hypomorphic, and thus not reconstructible. Every vertex-deleted subgraph of any of these graphs is isomorphic: they all are the disjoint union of an infinite number of copies of T.

9 The question of reconstructibility for locally finite infinite trees (the Harary–Schwenk–Scott conjecture from 1972) was a longstanding open problem until 2017, when a non-reconstructible tree of maximum degree 3 was found by Bowler et al. (see Bowler et al. 2017).

10 It will not be objected that "Digraph" is precisely the name of the journal and of the collection where Derrida published some of his works, and that the choice of this title could make sense: in truth, the absence of possibility of reconstructing the mathematical digraphs does not mean that this exception would be able to confirm Derrida's idea of "deconstruction" in the least. Moreover, Jacques Derrida is not known as a theoretician of graphs.

2

Philosophical Graphs

Let us first try to justify the consideration of graphs and combinatorics in our reflection: how closely are these structures related to philosophy?

It seems to us that the characteristic application of philosophy, which consists of constructing a global image of the world, presupposes, to be understandable, that the whole of reality is projected into some large classes of equivalence, in small numbers, so that they are easily remembered.

2.1. Ancient mappings

From this point of view, one of the simplest projections was that of the ancient Chinese, for whom the world is summed up in the Boolean opposition of yin and yang, to which, it is true, the system of five agents must be added (see Figure 2.1).

In fact, the yin-yang mapping is more complex than it seems: the yin is present in the yang and vice versa, so that the construction goes beyond the simple opposites white-black, be-nothing or zero-one. The mapping is Boolean, but it assumes, at its base, not a simple Boolean set $B = \{0, 1\}$ containing only two elements, like in Parmenides, but an expansion of it, in fact, a Cartesian product of the simple Boolean set. With the Yi-King adjunction, the ancient Chinese had a more refined explanation of the world, using the $2^8 = 64$ combinations of a Boolean set constructed on the opposition continuous/discrete, applied to lines.

At the origin of this, there is a curious legend associating the Luo Shu square, the Yellow river map and the 8 trigrams of the Bagua invented by the hero

1 For a color version of all the figures in this chapter, see http://www.iste.co.uk/parrochia/graphs.zip.

Fu Xi (see Figure 2.2). After a huge deluge in ancient China, the people offered sacrifices to the god of one of the flooding rivers, the Luo river, to try to calm his anger. A magical turtle emerged from the water with the curiously unnatural Lo Shu pattern on its shell: an unconnected graph representing the integers 1 through 9, which can also be arranged in a three-by-three grid – in fact, a magic square where the sums in each of the three rows, in each of the three columns and in both diagonals, add up to 15 (the number of days in each of the 24 cycles of the Chinese solar year).

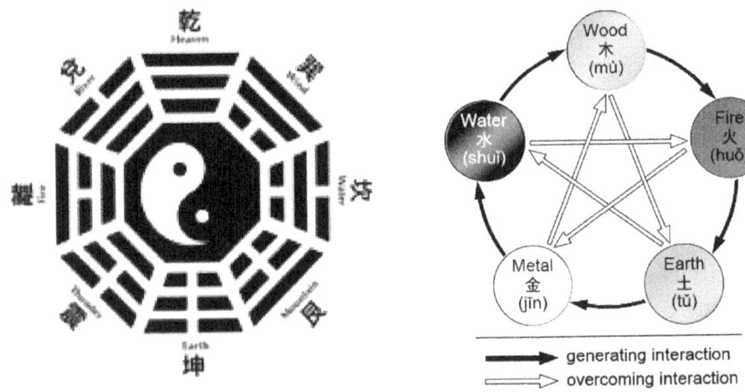

Figure 2.1. *The binary mapping of yin-yang (source: right: https://bmataichidata.com/five-element-theory-page-1.php; left: modified from: https://www.nationsonline.org/oneworld/Chinese_Customs/bagua.htm)*

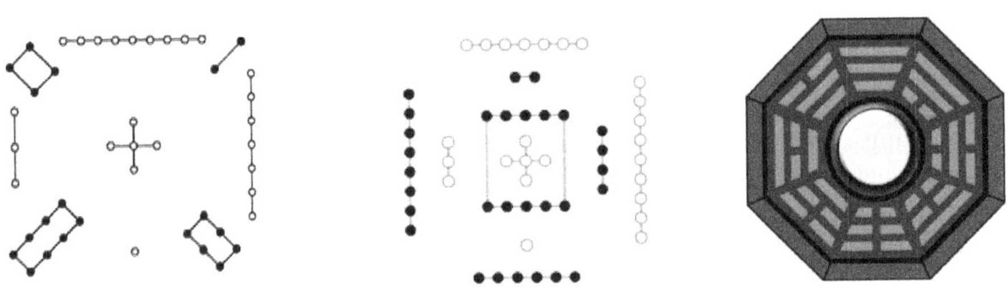

Figure 2.2. *Luo Shu square, yellow river map and the Bagua (source: left: Wikipedia article: Lu shuo square (the luòshū 洛書); center and right: Wikipedia article: Yellow River map)*

The Lo Shu is connected with the "yellow river map", sometimes interpreted as a means of controlling the yellow river and its tributaries (see Figure 2.3), which, many times in the past, changed its course, settling in new beds, often accompanied by death and devastation to the human population.

From the yellow river map, Fu Xi is supposed to have derived the first eight trigrams (or Bagua) which later composed the Yi-King.

We still find other graphic figures in ancient China, such as the system of subdivision of the sky into 28 lunar compartments, that of the 24 cardinal points and 12 terrestrial branches (the latter being used to measure the time) or the complicated Chinese geomantic compass or Feng Shui compass, the Feng Shui being the art of locating prosperous places (see Figure 2.4).

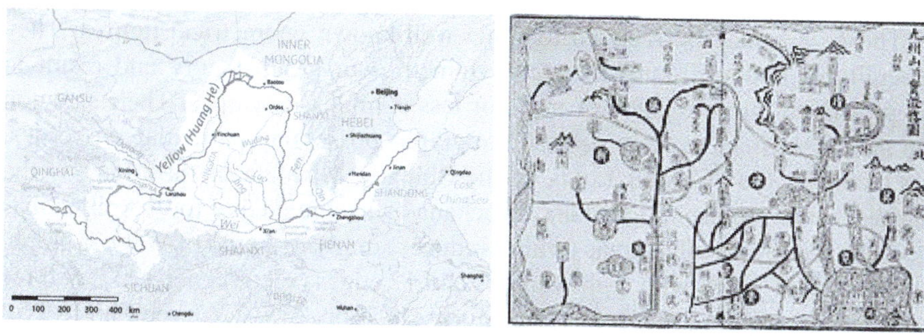

Figure 2.3. *The yellow river and a depiction of the system of the nine zhou (or river-islands) (source: left: Wikipedia article: Yellow River; right: Wikipedia article: Yellow River map)*

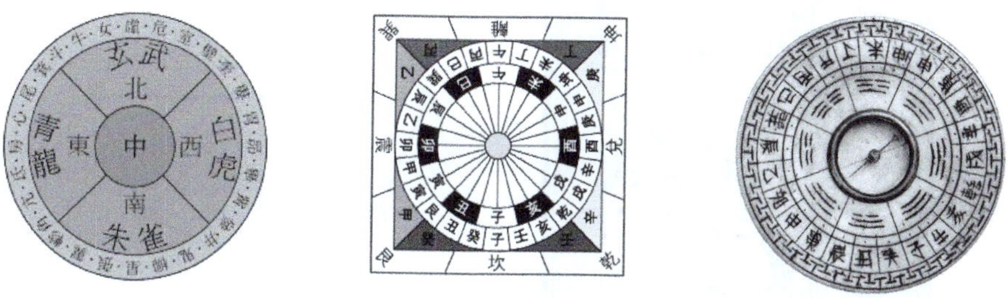

Figure 2.4. *Subdivision of sky, earth and Fenshui compass (source: left: Wikimedia Commons: https://commons.wikimedia.org/wiki/User:Rocket000/SVGs/%2A; center: Wikipedia article: Kompass; right: https://www.dreamstime.com/royalty-free-stock-photos-feng-shui-compass-isolated-image29272608#res170862)*

In other cultures, the projection of the world into lines or figures has taken other forms, sometimes closer to art. For Hegel, they are not considered philosophy because they do not make use of real concepts. However, those figures have underlying structures of quasi-mathematical forms.

Let us take a glance at what Indian people call "mandalas".

Historically speaking, "mandala" is a Sanskrit term meaning "circle", and by extension, "sphere", "environment", "community". In Hinduism, according to experts in the field (see Herbert and Varenne 1985, p. 65), it refers to circular or circle-like figures, designates a structure or form of organization, or the corresponding drawing lying on the ground or on another support in connection with various rites. In Buddhism, it is mainly used for meditation. The diagram is usually full of symbols and can be associated with a deity. But some mandalas, very elaborate and codified, tend towards geometric abstraction.

These ones may be reduced to some well-known geometrical figures. They generally exploit well-known space symmetries in group theory and result in the projection into the plane of more or less complex polytopes. Their skeleton can also sometimes evocate some finite graphs possessing particular properties. For example, the "Central Lotus of the Shingon Mandala of the Matrix", the Garbhadhatu, representing Mahavairochana, surrounded by four Adibuddhas (golden) and four Bodhisattvas (white) reduces simply to a kind of Wagner graph, a three-regular graph with eight vertices and 12 edges (see Wagner 1937; and for a modern description, see Bondy and Murty 2007, pp. 275–276); see Figure 2.5.

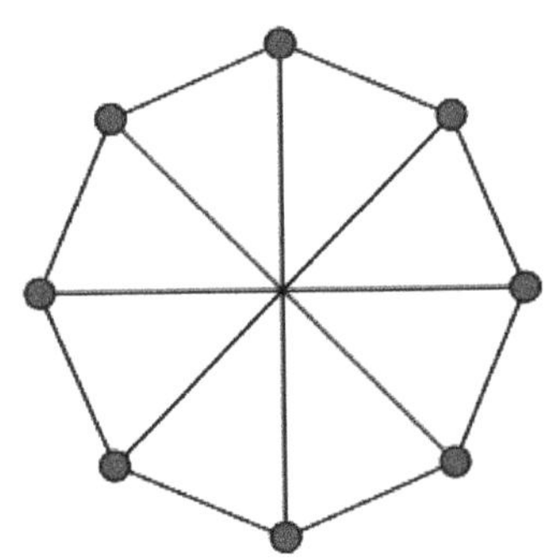

Figure 2.5. *A buddhist mandala and the Wagner graph (source: left: Wikipedia article: Mandala; right: Wikipedia article: Wagner graph)*

Of course, the Wagner graph arises through the theorem of Wagner (1937), stating that a finite graph is planar if and only if its minors include neither K_5 (the complete graph on five vertices) nor $K_{3,3}$ (the utility graph, a complete bipartite graph on six vertices). So, we could think that it has nothing in common with the buddhist mandala. Indeed, they both have the same structure.

2.2. Chinese tetragrams

The text *Taixuanjing* ("Canon of Supreme Mystery", Chinese: 太玄經, pinyin: *Tái Xuán Jīng*) was composed by the Confucian writer Yang Xiong (53 BCE-18 CE). The first draft of this work was completed in 2 BCE (in the decade before the fall of the Western Han dynasty). During the Jin dynasty, an otherwise unknown person named Fan Wang (Chinese: 范望) salvaged the text and wrote a commentary on it, from which our text survives today.

The *Taixuanjing* is a divinatory text similar to, and inspired by, the *Yi-King* (Chinese: 易經, pinyin: *Yijing*). Where the *Yi-King* is based on 64 binary hexagrams (sequences of six horizontal lines, each of which may be broken or unbroken), the *Taixuanjing* employs 81 ternary tetragrams (sequences of four lines, each of which may be unbroken, broken once or broken twice). Like the *Yi-King* it may be consulted as an oracle by casting yarrow stalks or a six-faced die to generate numbers which define the lines of a tetragram, which can then be looked up in the text. A tetragram drawn without moving lines refers to the tetragram description, while a tetragram drawn with moving lines refers to the specific lines.

The monograms are:

The unbroken line (–) for heaven (Chinese: 天; pinyin: *tiān*), once broken line (– –) for earth (Chinese: 地; pinyin: *dì*), twice broken line (– – –) for man (Chinese: 人; pinyin: *rén*). Numerically, the symbols can be counted as $- = 0$, $-- = 1$, $--- = 2$, and grouped into sets of four to count from 0 to 80. This is clearly intentional as a passage from Chapter 8 of the *Taixuanjing* points out the principle of carrying and place value.

2.3. Pythagorism and pentagram

The most trivial arrangements of lines are the simplest geometric configurations: line and points, triangles, squares, polygons of n sides, circles or a combination of these.

Among them is the pentagram, a figure of a five-pointed star close to the pentagon and usually made of alternate points connected by a continuous line. It appears in the earliest writing of Mesopotamia (precuneiform pictographic writing), c. 3,000 BCE, as the Sumerian sign UB. Its meaning in the cuneiform period (by 2,600 BCE) seems to be a Heavenly Quarter and also the four directions (forward, backward, left, right); the fifth direction was "above". The four directions corresponded to the planets Jupiter, Mercury, Mars and Saturn, with Venus as the Queen of Heaven (Schekina) above. These are the "smaller planets" (omitting the Sun and the Moon). Ishtar (Venus) was represented by the eight-rayed star (Elam).

Pythagoras may have become acquainted with the Pentagram during his sojourns in Egypt and Babylon (perhaps 554–533 BCE). However, the Pythagoreans used it as a sign of recognition.

The Pythagoreans apparently labeled the points or angles of the Pentagram with the Greek letters Υ, Γ, I, E, I, A (UGEIA), with EI standing for only one letter (see Figure 2.6). They also refer to elements, some organic cycle, the tetractys, the planets of the solar system known at the time and many other significations like functions of the self or metals, which will later still leave traces in the alchemy treatises. It sometimes appears in a pentagon and/or includes in itself a stellar pentagon.

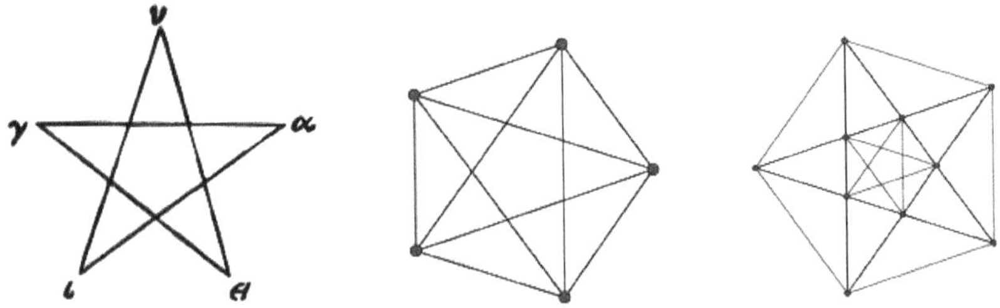

Figure 2.6. *The pythagorean pentagon*

But, indeed, Pythagoreans were above all interested in the properties of this figure and especially the ratios between the segments, which showed remarkable proportions (especially the golden ratio) and revealed the harmony of the universe to them.

Plato, whose education was a pythagorean one, kept some of these beliefs in the *Timaeus* and generalized the ancient Pythagorism in his own representation of the world, the famous schema of the Line in the *Republic*, a segment divided and subdivided in the same ratio, not precise, but including, of course, as a particular case, the golden ratio.

In astronomy, the pentagram was later to be associated with the apparent movements of the planet Venus seen from the Earth, which is sometimes poetically called the "dance of Venus" or, more mathematically, the "pentagram of Venus". This pentagram corresponds to the fact that five "lobes" exist where

Venus makes its closest approach to Earth. At each closest approach, the planet moves backwards compared to its usual motion across the sky: this is called retrograde motion, giving Figure 2.7[1].

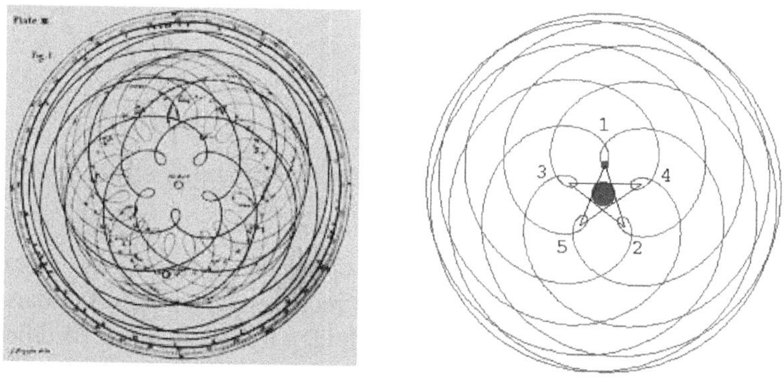

Figure 2.7. *The pentagram of Venus*

2.4. n-grams and some figures of the world

Since ancient times, as we can see, thinkers (philosophical or religious) have attempted to inscribe the world in simple figures, which summarize its main aspects (trigrams of Fuxi, combined in hexagrams, tetragrams of Yang Xiong, Pythagorean pentagrams, etc.). An interesting question now is how far did these attempts go and what more complex figures (if any) did they use.

It seems that the hexagram, the heptagram and the octagram (especially in their regular and starry form – see Figure 2.8) were retained by several religions, which overinvested them.

It also seems that, like the triangle, circle or square, the hexagram has been created by various people with no necessary connection to one another. Six-pointed stars have been found in cosmological diagrams in Hinduism, Buddhism and Jainism with different meanings.

1 Actually, this is only approximately true. The Earth orbits the Sun once every 365.256 days. Venus orbits the Sun once every 224.701 days. So, Venus orbits the Sun in $224.701/365.256 \approx 0.615187$ Earth years. Now, we know that $8/13 \approx 0.615385$. So, in eight Earth years, Venus goes around the Sun almost 13 times. During this eight-year cycle, Venus gets as close as possible to the Earth about 13 - 8 = 5 times. And so, each time it does, Venus moves to a new lobe of an approximative pentagram of Venus. The new lobe is 8 - 5 = 3 steps ahead of the last one. See Ferguson (1799).

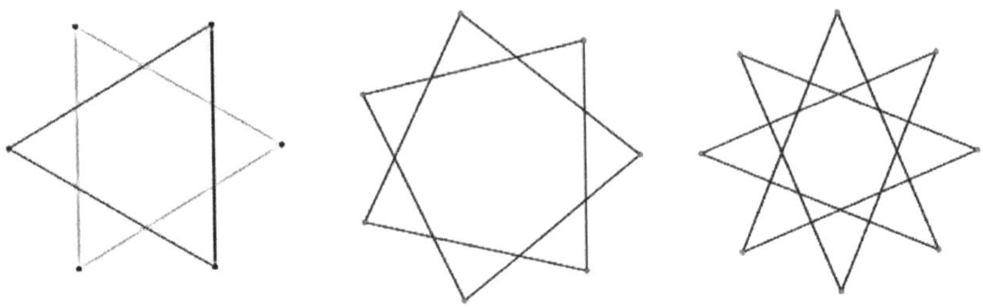

Figure 2.8. *Regular hexagram, heptagram and octagram*

But, of course, we find the main use of the hexagram in the Abrahamic religions. The Star of David is a generally recognized symbol of Judaism and Jewish identity. Its usage as a sign of Jewish identity began in the Middle Ages, though its religious usage began earlier (it apparently appears on a stone bearing the shield from the arch of a 3rd to 4th century synagogue in the Galilee). It may be found in some churches, but has also been used by Muslims and is present in some mosques.

This very widespread symbol in various fields (heraldry, theosophy, occultism, Rosicrucian and Hermetic Magic, Freemasonry) is known to people of the north (the Ulster Banner flag of Northern Ireland) and people of the south (a Flag used by rebels during the Whiskey Insurrection in South-Western Pennsylvania, 1794). Botanists used it (there is a plant named Solomon's seal (Polygonatum multiflorum) in the lily family), as well as alchemists (in alchemy, the two triangles represent the reconciliation of the opposites of fire and water).

So, all of this comes from the Greek and, via pre-Muslim Arab literature, arrives in Alchemy, where the superposed triangles thus represent combinations of elements. This is also the case for the svastika, which is not a hebrew symbol (see Scholem 1949).

Concerning the heptagram, we know it was used in Christianity to symbolize the seven days of creation and became a traditional symbol for warding off evil, but the symbol was already used in Kabbalist Judaism. In Islam, the heptagram is supposed to represent the first seven verses in the Quran. It is also known among neopagans as the Elven Star or Fairy Star and treated as a sacred symbol in various modern pagan and witchcraft traditions. In alchemy, a seven-sided star can refer to the seven planets which were known to early alchemists. It is also very widespread, even in Polynesia.

The octagram itself may have an islamic (Rub el Hizb) or semitic (the ancient Sumerian goddess Inanna and her East Semitic counterpart Ishtar) character. Used in Indonesia to represent the Hindu gods of the directions, it probably inherits its usage in compasses to represent the cardinal directions for the eight principal winds from there. But it is also widely used, from Latvians to Mapuche, including Norwegian iconography.

Naturally, for each of these figures, there are many equivalents, regarding, in particular, their method of construction. The octagram, for example, can be seen as the isogonal truncation of a square, as a two regular octagrammic star compound, the first constructed as two squares and second as four degenerate digons[2]. There are other isogonal and isotoxal compounds including rectangular and rhombic forms (see Figure 2.9).

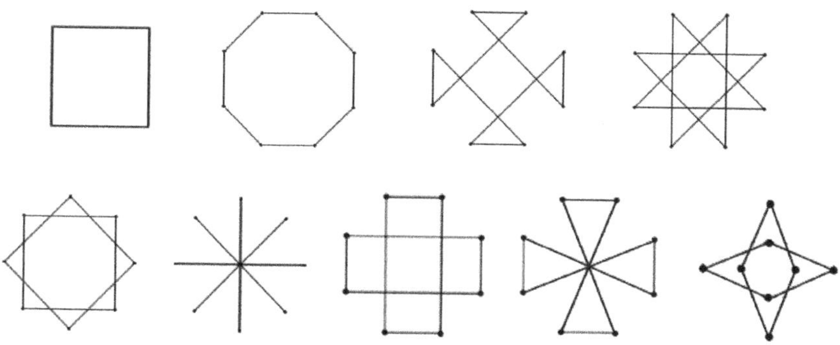

Figure 2.9. *Different forms of octagrams*

Now, mark nine points on a circle numbered 1,2,3,4,5,6,7,8, with 9 at the top. Then, divide the numbers 1 through 6 by 7 for the "Law of Sevenfoldness" yielding 1/7, 2/7, 3/7, 4/7, 5/7, 6/7 giving: .142857142857142857..., .285714285714..., .428571..., .571428..., .714285..., .857142 ... Note that these results contain the same repeating decimal. Draw a line connecting these numbers on the circle (note again that the repeating decimal numbers when added together equal 0 mod 9 ($1 + 4 + 2 + 8 + 5 + 7 = 27 = 9 \times 3$)). The missing numbers 3, 6 and 9 can then be connected to form an inner triangle with its apex at the top. Thus, the famous symbol of the enneagram appears.

It was in 1916 that the mystical philosopher Georges Ivanovitch Gurdjieff introduced the enneagram (see Figure 2.10), a graph borrowed from a

2 In geometry, a digon is a degenerate polygon with two sides (edges) and two vertices.

hypothetical Sarmouni brotherhood[3] in Central Asia, in connection with his method for raising awareness called the Fourth Way.

However, the first publication on the subject dates back to the publication of Ouspensky's book, *Fragments of an Unknown Teaching* in 1949. It is Ouspensky who grafted onto this diagram a system of equivalences with the functions of the human body and the planets. Then, Rodney Collin-Smith, one of his disciples, made the enneagram corresponding to the course of a human life, which he divided into nine periods and different psychological types. This last correspondence found a resurgence of topicality in 1970 with Oscar Ichazo and Claudio Naranjo. The enneagram then became a typology describing nine types (or enneatypes): 1. The Reformer; 2. The Helper; 3. The Achiever; 4. The Individualist; 5. The Investigator; 6. The Loyalist; 7. The Enthusiast; 8. The Challenger; 9. The Peacemaker. They correspond to nine fundamental motivations, themselves linked to nine fundamental avoidances. However, according to some of his disciples, Gurdjieff never used the enneagram as a tool for describing personality types.

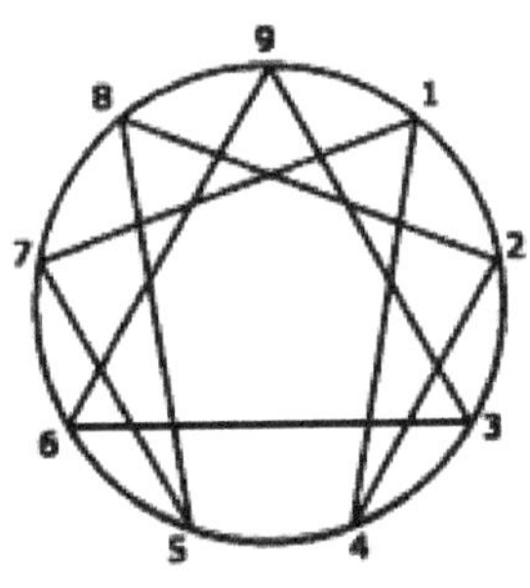

Figure 2.10. *The figure of an enneagram*

According to Gurdjieff, the enneagram would show two fundamental laws of the world: the "Law of Seven" and the "Law of Three", which are, respectively, a law of octaves, according to which phenomena evolves in seven steps, and a Law of Three, according to which phenomena are produced by three forces, laws presented by Gurdjieff as global laws appearing on all scales of the world and essential to his cosmology. This cosmology, derived from alchemy and more ancient sources, is supposed to complement or complete modern experimental science. For Gurdjieff, everything is material, including consciousness and spirit,

3 He could also, according to some, of Kaballistic, Pythagorean, hermetic or even Sufi origins, remember the Leonhard Euler's nine-point circle or even the "nonagram" of Ramon Lull.

and all matter can be assigned a “density”, consciousness and spirit, for example, having lower “densities” (and therefore higher “vibrations”) than, for example, water or, a fortiori, rock. Of course, this theory, like the enneagram, has no real justification.

Let us now rapidly explore a subset of the zoo of graphs for, say, $n < 20$.

Let us begin with the decagram. One of its main realizations in judaic culture is the sephirotic tree (also known as the Tree of Life or Jacob’s Ladder). Its structure is shown in Figure 2.11.

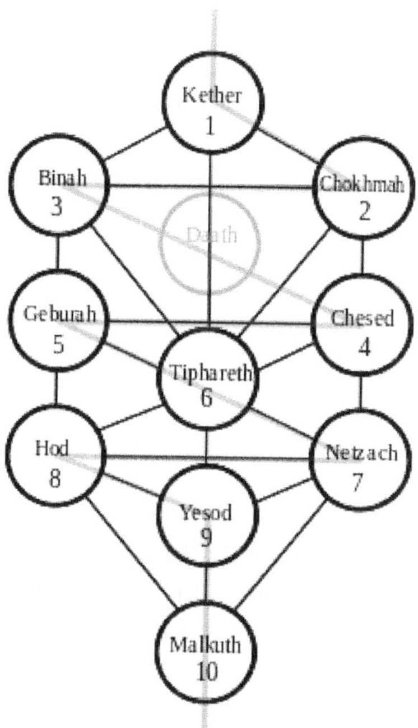

Figure 2.11. *Instantiation of a decagram: the Tree of Life*

As shown in Figure 2.11, the divisions of the tree consist of:

1) Keter – crown: divine will;

2) Chochmah – wisdom;

3) Binah – understanding;

4) Daat – knowledge (central state of unity of the 10 Sephirot);

5) Chesed – kindness;

6) Gevurah – severity;

7) Tiferet – beauty;

8) Netzach – victory;

9) Hod – splendor;

10) Yesod – foundation;

11) Malchut – kingship.

We will not enter into a commentary on this "tree": it would be necessary to mobilize a whole hermeneutic. Let us just say that, like any summary of the world, it mixes different realities. It seems that we can identify in it metaphysical, moral and psychological meaning. Metaphysically, the tree of life can be seen as the representation of the process of creation, putting to work, both in the Macrocosm that is the Universe and in the Microcosm that is the Human Being, creative energies or powers from the Creator. Morally, the Creator is a guide that must make it possible to rise, by a progressive initiation and a kind of individual re-creation, from the material world towards spiritual realities, by the crossing of different "veils" giving access to hidden realities. Finally, it explains real existence and individual behavior: according to Kabbalah, everyone has within them, a percentage of each sephiroth, which explains their psychology.

The figures with 11 (hundecagram) and 12 (dodecagram) sides were rather architectural or societal symbols. Regular tridecagrams have been used in numismatics as models of gold coins. Just like regular tetradecagrams, pentadecagram and hexadecagrams were used in art. A regular heptadecagon is a heptadecagon whose 17 sides have the same length and whose internal angles have the same measurement. They have seven stars (the heptadecagrams noted $\{17/k\}$ for k from 2 to 8) and one convex (noted $\{17\}$). This is the latter when we speak of a "regular heptadecagon".

The announcement of the construction with the rule and the compass of the regular heptadecagon was made by Carl Friedrich Gauss in 1796, in a short article, "Neue Entdeckungen".

A regular octadecagon is an octadecagon whose 18 sides have the same length and whose internal angles have the same measurement. They have two stars (the octadecagrams noted 18/5 and 18/7) and one convex (noted 18). This is the latter when we speak of a "regular octadecagon".

But nothing seems to be associated with the octadecagrams resulting from it, anymore than with the ennedecagon or the corresponding enneadecagrams.

Of course, many other figures may be, in theory, constructed. For 30, 40, 50... sides, we have a triacontagon, tetracontagon, pentacontagon, etc. A polygon with 100 sides is a hectagon, with 1,000 sides a chiliagon, with 10,000 sides a myriagon, with 100,000 sides a decakis myriagon, with 1,000,000 sides a megagon, etc.

But if figures may help our understanding, as Descartes said, there are some limits to imagination:

> I cannot in any way imagine the thousand sides of a chiliagon [as I do the three sides of a triangle], nor do I, so to speak, regard them as present [with the eyes of my mind]. And although in accordance with the habit I have formed of always employing the aid of my imagination when I think of corporeal things, it may happen that in imagining a chiliagon I confusedly represent to myself some figure, yet it is very evident that this figure is not a chiliagon, since it in no way differs from that which I represent to myself when I think of a myriagon or any other many-sided figure; nor does it serve my purpose in discovering the properties which go to form the distinction between a chiliagon and other polygons. But if the question turns upon a pentagon, it is quite true that I can conceive its figure as well as that of a chiliagon without the help of my imagination; but I can also imagine it by applying the attention of my mind to each of its five sides, and at the same time to the space which they enclose (Descartes 2016, VI).

Moreover, it is only the figures comprising few points or sides which are generally useful as a support for reflection and which generally serve to build the architectonics of philosophical systems.

2.5. Graphs and classical systematicity

Philosophical systematicity has taken different forms in the course of history. The philosophical doctrines of Antiquity, although they have a coherence, do not take this one as an object of reflection. The Platonic Dialogues or the Aristotelian Encyclopedia are not intended to be systems. The same can be said of medieval "sums", which have a taxonomic structuration and resemble classifications, due to pedago. In the 17th century, however, coherent thought tends to identify itself with deductive thought, the model of Euclidean geometry having had a considerable influence on the thinkers of the time. A demand for efficiency is emerging. So, a philosophical system becomes a finite complex of statements whose purpose is an optimal representation of reality, in a way that integrates the main aspects.

The idea of an optimal representation of reality can now be expressed as follows: given a set of objects or events U, named "universe", find in U a subset S which is the best selection possible, given certain constraints. In other words:

$$\text{Maximize:} \quad z = cx,$$

$$\text{under the constraints} \begin{cases} Ax \leq b, \\ x_j = 0 \; or \; 1, \; j = 1 \text{ to } n \end{cases}$$

where A is an $m \times n$ matrix with non-negative integer coefficients, b is an m-vector of positive integers and c is an n-vector of non-negative integers. When $m = 1$, the problem is known as the "Knapsack problem". Indeed, a hiker must choose, from a list of objects to carry in their bag, a subset whose weight does not exceed a given weight and which is the best selection possible. b represents the maximum weight they can carry, a_j is the weight of the object j and c_j is a measure of its "utility". In the case of philosophical systems, there are many constraints that are difficult to evaluate and which concern both the length of the utterances or the total text itself and the number of conceptual categories or the various "syntheses" it contains, etc.

The interaction of different statements can be manifested by highlighting a particular relationship. Thus, the relation of deduction describes the interaction of the various axioms, definitions and propositions of Spinoza's *Ethics*, which form a graph of a certain complexity, but whose matrix can be easily drawn up and the properties studied. In the graph of Figure 2.12, the arrows mean "serves to demonstrate".

In Table 2.1, "P" stands for "Proposition", "D" for definition, "A" for "Axiom", "CPn" for "corollary of the proposition Pn". When one statement serves to demonstrate another one, we put 1 at the intersection of the line and the corresponding column. Otherwise, we put 0. For example, the definition D3 serves to demonstrate the propositions P1, P2, P4, P5, P6 and the corollary of the proposition P6. We leave aside the two scholias of Proposition 8, which are not proven by anything and do not participate in the deduction. So, we get the matrix of the deduction's graph for the first nine propositions of *Ethics I*.

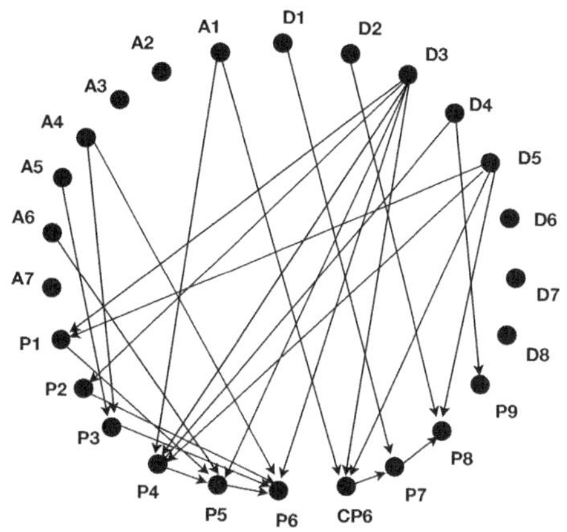

Figure 2.12. *The graph of deduction for the first propositions of Spinoza's Ethics I*

○	P1	P2	P3	P4	P5	P6	CP6	P7	P8	P9
D1	0	0	0	0	0	0	0	1	0	0
D2	0	0	0	0	0	0	0	0	1	0
D3	1	1	0	1	1	1	1	0	0	0
D4	0	0	0	1	0	0	0	0	0	1
D5	1	0	0	1	0	0	1	0	1	0
D6	0	0	0	0	0	0	0	0	0	0
D7	0	0	0	0	0	0	0	0	0	0
D8	0	0	0	0	0	0	0	0	0	0
A1	0	0	0	1	0	0	1	0	0	0
A2	0	0	0	0	0	0	0	0	0	0
A3	0	0	0	0	0	0	0	0	0	0
A4	0	0	1	0	0	1	0	0	0	0
A5	0	0	1	0	0	0	0	0	0	0
A6	0	0	0	0	1	0	0	0	0	0
A7	0	0	0	0	0	0	0	0	0	0
P1	0	0	0	0	1	0	0	0	0	0
P2	0	0	0	0	0	1	0	0	0	0
P3	0	0	0	0	0	1	0	0	0	0
P4	0	0	0	0	1	0	0	0	0	0
P5	0	0	0	0	0	1	0	0	0	0
P6	0	0	0	0	0	0	0	0	0	0
CP6	0	0	0	0	0	0	0	1	0	0
P7	0	0	0	0	0	0	0	0	1	0
P8	0	0	0	0	0	0	0	0	0	0
P9	0	0	0	0	0	0	0	0	0	0

Table 2.1. *The relation of deduction in Spinoza's Ethics (first propositions)*

Despite the partial nature of this matrix (or the corresponding graph), we can still make some remarks immediately. Some definitions or axioms have no use in this part of the *Ethics*. These are, for example, the definitions D6, D7 and D8, or alternatively axioms A2 and A7. Some propositions are demonstrated using several elements (definitions, axioms or previous propositions), such as P1, P4, P5, P6 and CP6. Others seem more fragile: for example, P7 is only demonstrated using definition 1 and P9 is only demonstrated using definition 4.

In some systems, the interaction between utterances of the same linguistic level can have remarkable properties: thus, in Hegel, the conceptual tree of the *Encyclopedia* (see Figure 2.13) has the following form.

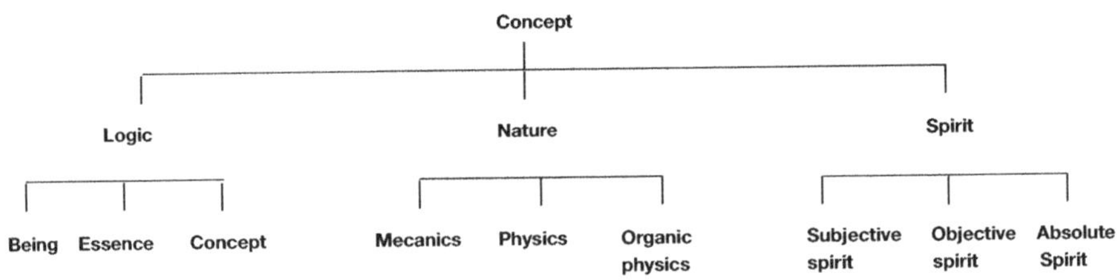

Figure 2.13. *The main structure of Hegel's Encyclopedia of philosophical sciences*

In Wittgenstein, the *Tractatus logico-philosophicus*'s structure of propositions combines both a multilevel hierarchy and a vector organization.

Recall here the following definition:

DEFINITION 2.1 (VECTOR LATTICE).– *Let n be sets $A, B, ..., S$, each totally ordered by a relation $<$. Then, the Cartesian product $A \times B \times ... \times S$ is ordered and forms a lattice called a "vector lattice", which is reduced to a total order – that of the dictionary. We consider, in fact, the following domination relation: an n-uple $(A_i, B_j, ...S_t)$ will dominate an n-uple $A_{i'}, B_{j'}, ...S_{t'})$ if the first r elements (starting arbitrarily from the left) of the two n-tuples are equal, but the $(r + 1)$ th element of the first is superior (for the relation of order which concerns it) to the $(r + 1)$ th element of the second. So, we have a total order.*

Example: The two lattices (a) and (b) in Figure 2.14 are vector lattices.

In the *Tractatus*, the numeration of propositions corresponds to a lexicographical order: 1.11 precedes 1.12; 2.011 precedes 2.012, etc.

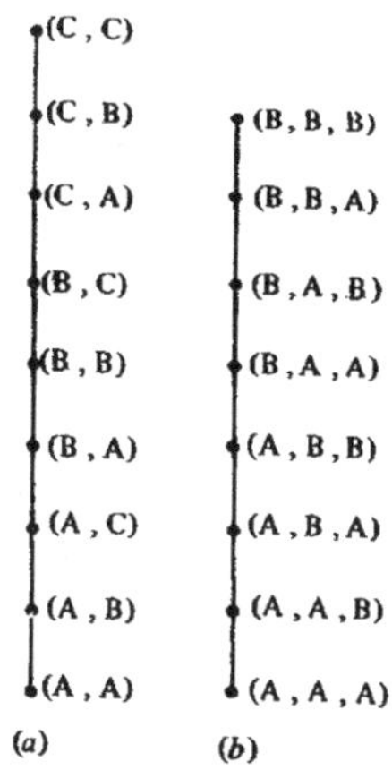

Figure 2.14. *Two examples of vector lattices*

That said, this apparent structure may conceal another one, slightly more complex, well highlighted by Granger (1969). To each of the main propositions of the *Tractatus*, observes Granger, subordinate observations dominate themselves from other remarks, and this hierarchy is materialized in the numbering system adopted by the author. Thus, for the first proposition, which has only three levels, we have the following diagram[4]:

1

1.1 1.2

1.11 1.12 1.13 1.21

The absence of a level results in the presence of a zero in the virtual place. Thus, for the group of proposition 2:

2

——

.

——

2.01.

4 The numbers 1 and 2 refer to the first and the second proposition of the Wittgenstein Tractatus. The numbers 1.1, 1.2, 2.01 etc refer to the observations and remarks which develop these propositions.

Every proposition, wrote Granger, "has a definite level of importance and significance, but not its place in a chain. It seems that a relation of 'distance' is substituted to the relation of principle to consequence, which induces, on the set of propositions subordinated to a mother-proposal, a very extensive stratification, sometimes including up to six levels (...) We use the word 'distance' and not simply 'order relation'". This is because Wittgenstein wanted to note the place – empty – of levels not represented by propositions, so that each remark is affected by an index indicating the number of degrees which separate it from the mother-proposal. Thus, proposition 3 is immediately followed by a remark of the third degree – 3.001 – without any intermediate proposition (of the second or first degree, which would be noted 3.1 and 3.11) to serve as immediate support. Here is the beginning of diagram 3.

3

. 3.1

3.01 3.02 3.03 3.11

3.001

Instead of the total order of the lexicographical vector lattice, we would thus have a partial order, a kind of hierarchical classification rather analogous to that of the Hegelian system, except that it would not have its regularity and would relate to the sentences instead of concepts.

These partial orders, like, a fortiori, the total order of the Cartesian chain of reasons, remain, however, the exception.

In general, from the 17th century, a philosophical system has a network structure instead. It is even made of the intercrossing of a multiplicity of graphs, which are superimposed on each other, according to the semantic level considered. Thus, the Leibnizian image of the system, in this faithful to its Stoic origin as to the etymology of the word "sustema", adds to the idea of a methodical progression, that of an organic totality (see Goldschmidt 1953; Serres 1968). Descartes himself, who claims to be attentive to the order and connection of his reasons "rationum mearum seriem and nexum comprehendere" (*Metaphysical Meditations*, preface), seems to subordinate it to the "coherentia" or continuation and connection of the ideas. And this one is "such that for the proof (probationem) of each question (rei), all the things which precede it contribute to it and a large part of those which follow it" (Descartes, *Replies to the Fifth Objections*, V1; see

Beyssade 1979). The "logic" of the presentation is thus always accompanied by an "architectonic", in the sense that Gueroult (1957) gives this term after Kant.

From the point of view of graph theory, this language amounts to identifying the system to a closed set of speculative categories, whose graph is a very complex interaction graph, and in which any path from one category of the system leads to another category of the system. The orientation of the graph then makes it possible locally to find the deductive moments of the step, when they exist. It also manifests, in some systems, the circular nature of the philosophical demonstration, which makes sure that the place where we arrive is, in the end, the place from which we had left, but that we retrieve any other perspective (see Ladrière 1972).

This circularity is explained by the fact that philosophical discourse, unlike a body of scientific doctrine, finds itself unable to start from an axiom, formulated or not, from a "consensus gentium" or fact, and must therefore justify its beginning, as well as each of its steps. According to Weil, it is not

> an accessory quality which it could as well be dispensed with, on the contrary, where it lacks, the thought has not reached maturity, and even if all the categories would be present there, a doctrine which does not return to its beginning would not have elaborated all these categories in their purity and would have stopped at one of the instantiations of the category of meaning (see Weil 1967, p. 440).

It is up to Hegel to have established circularity as the criterion of philosophy, thus marking, by his refusal of a philosophy of reflection, the end of an era of Western thought. Formally speaking, circularity results in the existence of a "Hamiltonian cycle" or eventually a "Eulerian cycle" on the system graph. Let us recall the definitions of these two kinds of circularity.

DEFINITION 2.2 (Hamiltonian cycle).– *A Hamiltonian cycle, also called a Hamiltonian circuit, is a graph cycle (i.e. closed loop) through a graph that visits each node exactly once. A graph possessing a Hamiltonian cycle is said to be a Hamiltonian graph. By convention, the singleton graph* K_1 *is considered to be Hamiltonian even though it does not possess a Hamiltonian cycle, while the connected graph on two nodes* K_2 *is not.*

We give some examples of Hamiltonian cycles in Figure 2.15.

A trivial philosophical example may be given with Weil's logic of philosophy (see Figure 2.16(a)) and its 18 conceptual categories, which are connected in the following order: Truth (1), Nonsense (2), True and False (3), Certitude (4), Discussion (5), Object (6), The Ego (7), God (8), Condition (9), Conscience (10), Understanding (11), Personality (12), Absolute (13), Work (14), The Finite (15), The Action (16), Sense (17), Wisdom (18).

Weil's system, thus made up of a single circle of notions, is quite trivial. We will not say as much of Hegel's system (see Figure 2.16(b)), which presents itself, explicitly, as a "circle of circles".

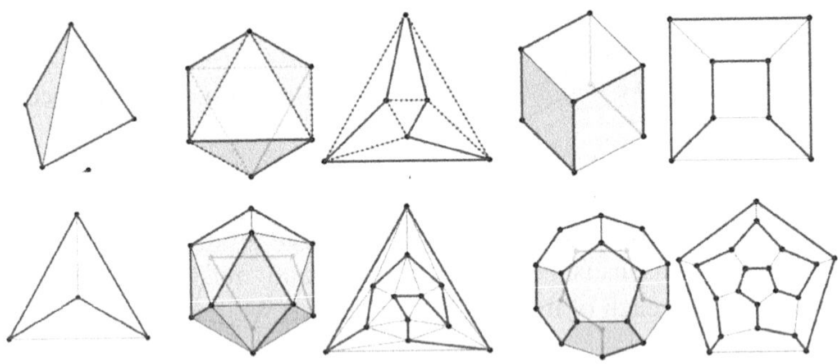

Figure 2.15. *Examples of Hamiltonian cycles in platonic solids (source: Wikipedia article: Hamilton path)*

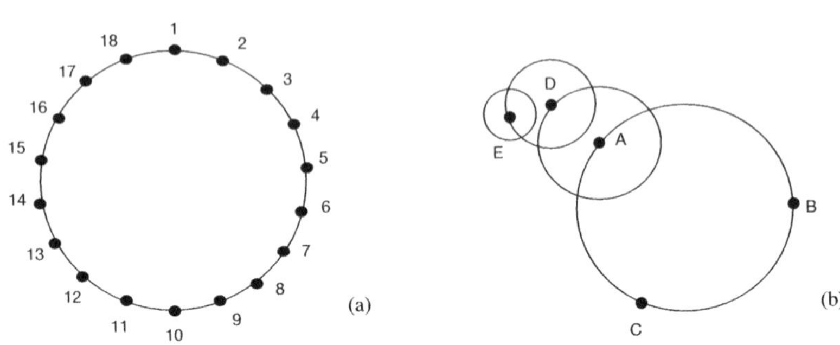

Figure 2.16. *Weil's circular system (a) and Hegelian epicycles (b)*

The three stages of the general movement of the concept associated with the Hegelian Encyclopedia (logic, philosophy of nature, philosophy of mind) – what we could qualify as a main circle – breaks down into a multiplicity of smaller circles, giving rise to decomposition, and so on.

Let us consider, for example, the structure of the Hegelian logic. Let A be "Being", whose automovement conducts to "Essence" and "Notion". A decomposes into D (Quality), whose automovement conducts to Quantity and to Measure. Now, D (Quality) decomposes itself into "Being" (E), whose automovement leads to "Determinate Being", and to "Being for self". Then, "Being", at this level, is decomposed into "Being as such" (G), whose automovement leads to "Nothing" and then to "Becoming". But we must not forget that, in its last development, "Quality" passes into "Quantity". So, as "Being for self" is the last development of "Quality", "Being for self" leads to "Quantity". In reality, all positions of the Hegelian logic (which are in fact only transient moments, just as the position of a stellar system's planet in astronomy) end up passing into another, so their denominations are only benchmarks. Thus, in a certain conceptual sphere, the development of a notion is the analog of the progress of a planet on an epicycle in the ancient Ptolemy system. For a planet whose the orbit would require two epicycles in addition to the deferent, we would get the exact Hegelian situation, with a dialectical self-movement composed of three phases (see Figure 2.17).

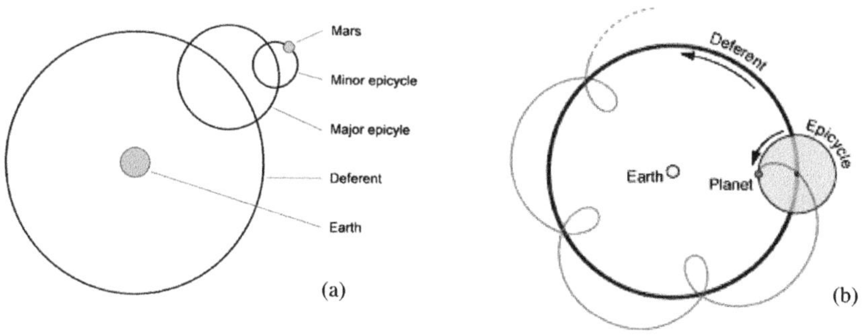

Figure 2.17. *Cycle and epicycle (a); movement of a planet (b)*

The main Hegelian circle is an analog of the *deferent* and the smaller ones are the analogs of the epicycles. Not only is this reminiscent of the geocentric Ptolemy's system exposed in the *Almagest*, but it is also, as has long been known, a perfectly rational way of breaking down a quasi-periodic movement into a multiplicity of periodic movements, which accomplishes, as we know, this magnificent mathematical operation called "the Fourier transform" (see Gallavotti 2001; Russo 2004, p. 91)[5]. In reality, Hegel, having perceived some kinds of

5 The interpretation of the Fourier transform in terms of deferent and epicyclical motions has been noted by many. The more ancient is probably in a memory from 1874 from Schiaparelli, reprinted in Schiaparelli (1926, p. 11). It should be stressed that the above "reduction" of a quasi-periodic motion to an epicyclic series is not unique.

periodicities in thought, achieves his system as a sort of philosophical discrete Fourier transform[6].

But we can still make the following remark: in epicycle-deferent astronomy, adding a second minor epicycle to account for observational discrepancies is mathematically equivalent to shifting the deferent into a so-called eccentric, or a circle with a center not at the Earth (see DeWitt 2010, p. 120), as already pointed out by Hipparchus. In other words, the addition of a second epicycle to the original epicycloid (compounded only of a deferent and a major epicycle) is supposed to have the same effect upon the movement as projected onto the outer circle of fixed stars by a line of sight from Earth, just shifting the original epicycloid by a distance equal to the diameter of the minor epicycle, as shown in Figure 2.18.

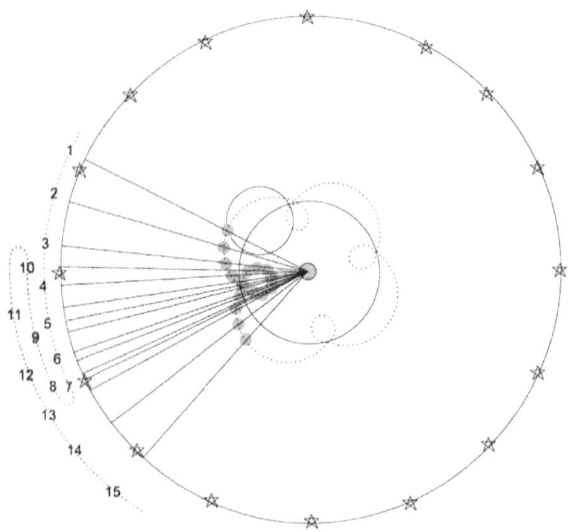

Figure 2.18. *Excentric equivalence*

This ensures that the Hegelian system, seen as the transposition of a Ptolemaic structure of the deferent-epicycles type, can also be expressed as a single circular chain satisfying the condition of Hamiltonian cyclicity.

6 Proof that the motions of the Ptolemaic lunar theories, as actually all quasi-periodic motions, can be interpreted in terms of epicycles, may be found in Gallavotti (2001). In addition, we also know that the discrete Fourier transform applies in graph theory, including on trees.

This formal translation makes it possible to ask many interesting questions: such a doctrine, which claims not to be a system, will be unmasked by the existence of a Hamiltonian circuit on its associated graph; we then ask what form the graph of a non-circular philosophical system can take; if the graph associated with the system is not Hamiltonian, we can determine under what conditions it can become associated (see the case of hypo-Hamiltonian graphs).

Let us now see what a Eulerian cycle is.

DEFINITION 2.3 (EULERIAN CYCLE).– *An Eulerian cycle, also called an Eulerian circuit, or Euler circuit, is a trail which starts and ends at the same graph vertex. In other words, it is a graph cycle which uses each graph edge exactly once. For technical reasons, Eulerian cycles are mathematically easier to study than Hamiltonian cycles.*

An Eulerian cycle for the octahedral graph is illustrated in Figure 2.19 with the platonic octahedron.

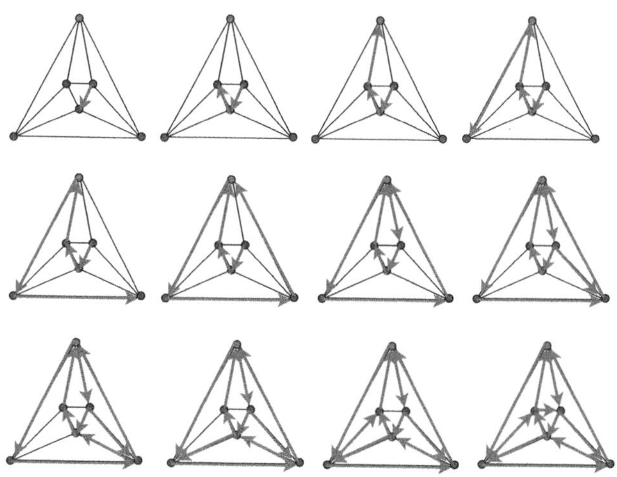

Figure 2.19. *Examples of an Eulerian cycle in the octahedron*

For philosophical systems, Hamilton circuits are more common than Eulerian circuits and easier to detect because it is more natural to identify the concepts with the vertices of a graph and their connections with its arrows than, for instance, to start with arrows, standing, for example, for statements or propositions, and to identify vertices with their interactions.

Relating philosophical systems to particular forms of graphs makes it possible to consider, without discrimination, a wide range of doctrines, all of different paces. It is a way of avoiding, as Heidegger (2021) wanted, the "conceptual confusion" which consists of assimilating the pseudo-systems of Plato and Aristotle, for example, to the pedagogical form of the medieval "sums" and the configuration taken by knowledge from the 17th century, which culminates, according to the author, in the "real systems" of Schelling and Hegel.

The present conception of systematicity makes it possible to maintain the word "system" instead of having to distinguish, as Heidegger does, the systematic thought proper from the simple "orientation towards the system", which introduces a teleology into the system. This latter expression, moreover, is too closely linked to the problem of the history of being, in other words, to Heidegger's own system or, if he does not accept this term, to the coherence of his own thought.

The idea of a conceptual interaction graph is therefore entirely neutral and does not prejudge the content of the system. As Gueroult (1979, p. 75) wanted, the system can be posited "as being, in its content, an authentic image of the original or conception of the impossibility of the image". Such a notion falls short of the author's opposition between "philosophy of truth" and "philosophy of existence", between "metaphysics of the cosmos" (based on external nature) and "metaphysics of the lived" (of which Christianity is the point of departure). It also precedes the formal opposition of the strongly architectured systems, with explicit methodology, and systems-fragments, with the secret resonances and interior (Presocratic words, maxims, aphorisms of romantic philosophers or Nietzschean thought).

2.6. Towards a new kind of systematicity

Of course, we may have the desire to get new forms of systematicity, and especially non-circular philosophical graphs. In the history of philosophy, some authors have tried to escape the Hegelian model in this way. A first solution is to build architectonics based on non-Hamiltonian or/and non-Eulerian graphs.

2.6.1. *Non-Hamiltonian and non-Eulerian philosophies*

Let us begin with some examples. Suppose some conceptual architecture is founded on the graph in Figure 2.20. It is obviously Eulerian but not Hamiltonian.

But a lot of graphs are Hamiltonian but non-Eulerian. See, for example, the graph in Figure 2.21.

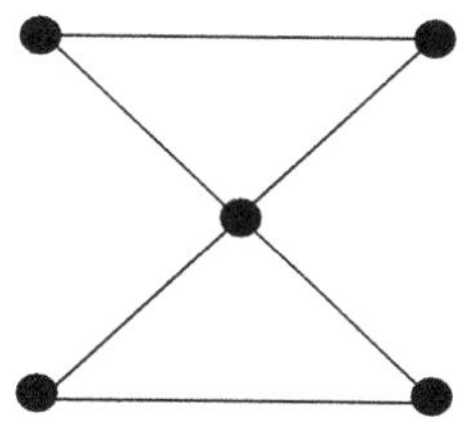

Figure 2.20. *A non-Hamiltonian graph*

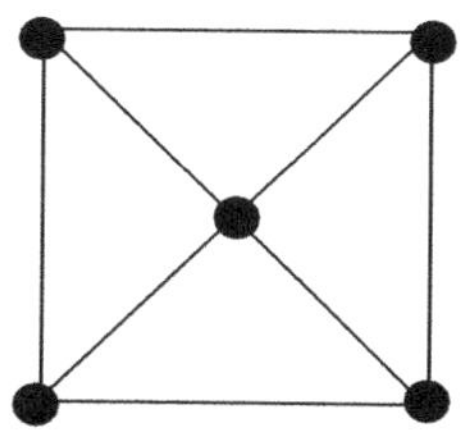

Figure 2.21. *A non-Eulerian graph*

Now, some graphs may be non-Hamiltonian and also non-Eulerian. See, for example, the graph in Figure 2.22.

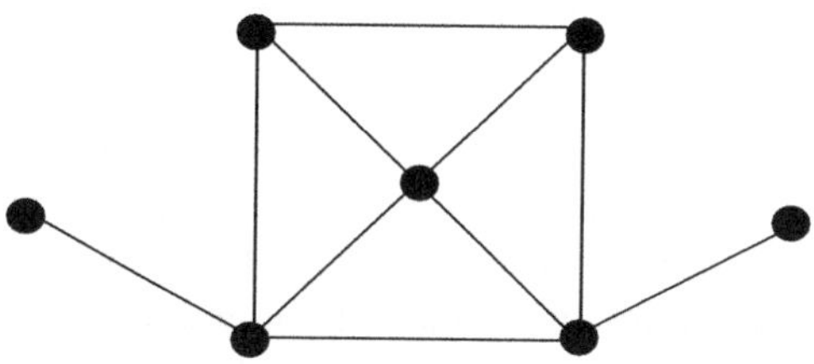

Figure 2.22. *A non-Hamiltonian and non-eulerian graph*

A simple way to escape Hamiltonian cycles is to take non-connected graphs. Among them, we find, for example, the hexagram, whose Schläfli symbol[7] is $\{6/2\}$, $2\{3\}$ or 3. It is a non-acyclic and non-bipartite claw-free graph with six vertices and six edges, a clique number 3, two triangles as components,

7 In geometry, the Schläfli symbol is a notation of the form $\{p, q, r, ...\}$ that defines regular polytopes and tessellations. It is named after the 19th-century Swiss mathematician Ludwig Schläfli (1814–1895), who generalized Euclidean geometry to more than three dimensions and discovered all its convex regular polytopes, including the six that occur in four dimensions.

matching number, maximum degree and minimum degree 2, chromatic index, chromatic number and circumference 3, algebraic, edge and vertex connectivity 0 and diameter infinity. The graph is shown in Figure 2.23.

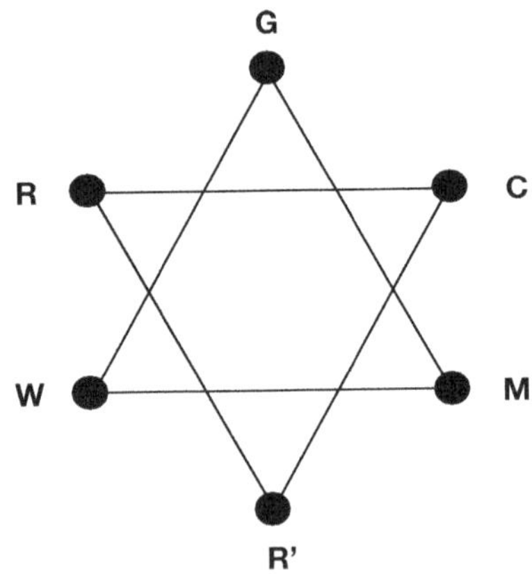

Figure 2.23. *Rosenzweig graph*

Such a graph has been associated with the philosophical doctrine of F. Rosenzweig, which the author presents in his book *Der Stern der Erlösung*. According to E. Levinas (preface to S. Moses, *System and Revelation*), the profound novelty of F. Rosenzweig

> lies in the contestation of the primordial character of a certain rationality: of those which illuminate the traditional philosophy "from the Ionian islands to Jena" – from the Presocratics to Hegel – which consisted in totalizing the natural and social experience, in disentangling them and linking them together with the categories, until they built a system that included the religious order itself.

In thinking of religion as "the original horizon of all meaning", Rosenzweig would have held to an exploded totality:

> the idea that everything is not assemblable: God, the World, Man – the three regions of the special ontology – do not assemble in one whole. Eclatment of the totality affirmed against Hegel. Of course. But also speculative gesture more general, being drawn by a radicalism that questions the most spontaneous and the most natural movement of philosophy: that which consists in embracing the thinkable, in elevating it to the universal in the unity of the genre, or integrating it among the moments of an ascending dialectic.

However, we ask ourselves a little question: what if this so-called thought of non-totalization and anti-systematicity proved to have constructed its doctrine in a basically traditional way, in any case not fundamentally different from the Hegelian way?

Mathematics leads to this question: the Rosenzweig hexagram is not Hamiltonian, of course, but a variant of it is indeed Hamiltonian: let us take a look at the following graph, in which we have reestablished links that Rosenzweig has obscured, but which are indeed present in his text (see Figure 2.24).

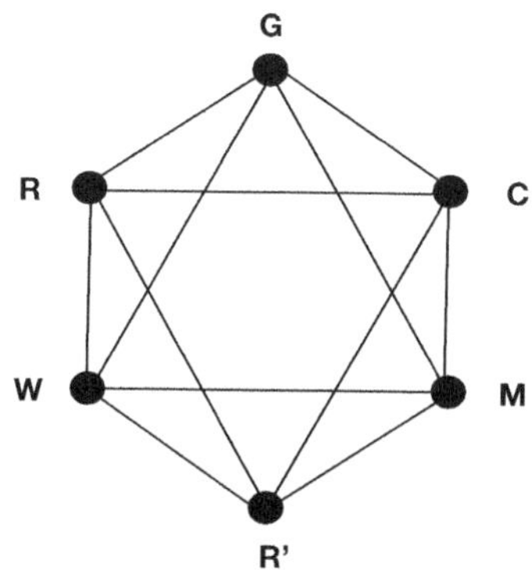

Figure 2.24. *A variant of the Rosenzweig graph*

This last graph is, in fact, also Eulerian. Moreover, note that, from the viewpoint of graph theory, no mystical sense is attached to this figure. We can easily see that the two following graphs are topologically equivalent and have the same properties (see Figure 2.25).

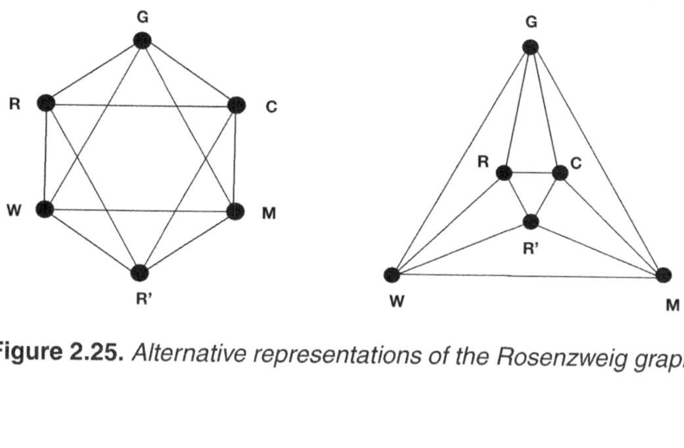

Figure 2.25. *Alternative representations of the Rosenzweig graph*

Now, take a more general point of view: do sufficient conditions for a graph to be non-Hamiltonian or non-Eulerian exist?

Consider first non-Hamiltonicity: a graph is non-Hamiltonian if it satisfies one of the following conditions (the list is not exhaustive):

1) it is not connected;

2) there exists a vertex which has degree <2;

3) it contains an induced subgraph in the form of a subdivision of a theta-graph.

To avoid the consequences of abusive totalization, the choice of Rosenzweig and of most contemporary philosophers is to refuse the condition (1), i.e. to disconnect certain parts of their system, which amounts, more or less, to abandoning systematicity or at least concealing it. Obviously, mathematics shows that there are other possible choices, probably less expensive.

There exist, in fact, a lot of *connected* non-Hamiltonian graphs. The numbers of connected simple non-Hamiltonian graphs on $n = 1, 2, \ldots$ nodes are $0, 1, 1, 3, 13, 64, 470, 4921, \ldots$, and the corresponding number of not-necessarily-connected simple non-Hamiltonian graphs on $n = 1, 2, \ldots$ nodes are 0, 2, 3, 8, 26, 108, 661, 6150, 97585, ... Among them are the graphs in Table 2.2.

Graph G	$\lvert V(G)\rvert$
Singleton graph	1
Claw graph	4
Theta-0 graph	7
Petersen graph	10
Herschel graph	11
No perfect matching graph	16
First Blanuša snark	18
Second Blanuša snark	18
Flower snark	20
Walther graph	25
Coxeter graph	28
Double star snark	30
Thomassen graph	34
Tutte's graph	46
Szekeres snark	50
Meredith graph	70

Table 2.2. *Some non-Hamiltonian graphs*

Philosophers are therefore spoiled for choice and may use very complex conceptual graphs for constructing their systematic organization of experience.

Now, consider the question of Eulerian cycles. What are the sufficient conditions for a graph to be non-Eulerian? The answer is well known since the Eulerian solution of Kœnigsberg's bridge problem. We know, from Euler (1736), the following theorem:

THEOREM 2.1.– *Let $X = (V, E)$ be a connected graph. Then, X is an Eulerian graph if and only if each vertex of X has an even degree.*

So, to get a non-Eulerian graph, it is sufficient that some vertex of the graph has an odd degree.

Now, if we combine this non-Eulerian condition with one of the non-Hamiltonian ones, we necessarily get graphs which are non-Hamiltonian and non-Eulerian.

2.6.2. *Pancyclic graphs and Metahegelianism*

But there exists another way to escape Hegelianism, which is not to forbid the existence of cycles or circuits in the conceptual graph of the system: it may be to intensify or generalize them. Look for instance at what is called a pancyclic graph.

DEFINITION 2.4 (PANCYCLIC GRAPHS).– *A graph G of order $n \geq 3$ is pancyclic if G contains all cycles of lengths from 3 to n. G is called vertex-pancyclic if each vertex v of G belongs to a cycle of every length ℓ, $3 \leq \ell \leq n$ (see Figure 2.26).*

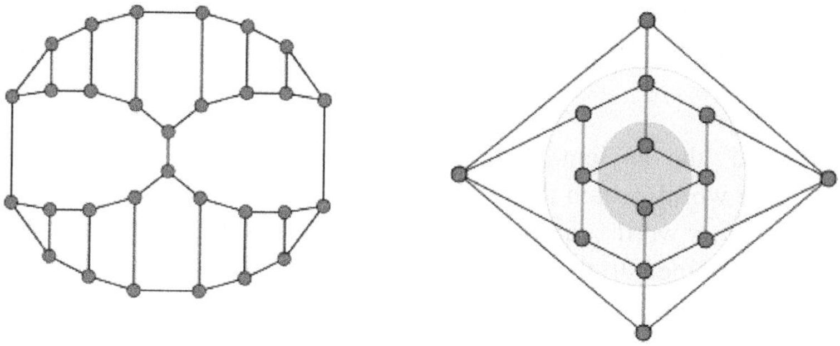

Figure 2.26. *Examples of pancyclic graphs (source: left: Wikipedia article: pancyclic graph; right: Wikipedia article: outplanar graph)*

This kind of construction may be associated with a sort of Metahegelianism. In such systems, the cycles can take any length $\ell \geq 3$, without linking or leading to a single cycle. The system contains cycles, but it is not a "circle of circles" in Hegel's sense.

Moreover, we can see that the pancyclic graph of Figure 2.26 (right) is in fact a three-outplanar graph. A k-outerplanar graph is a planar graph that has a planar embedding in which the vertices belong to at most k concentric layers. More formally, a graph is k-outerplanar if it has a planar embedding such that, for every vertex, there is an alternating sequence of at most k faces and k vertices of the embedding, starting with the unbounded face and ending with the vertex, in which each consecutive face and vertex are incident to each other.

Pancyclic graphs are a generalization of Hamiltonian graphs, which have a cycle of the maximum possible length. As a result, the maximal outerplanar graphs, which are also pancyclic, satisfy a stronger condition than if they were Hamiltonian: i.e. for each vertex s and for each k going from three to the number of vertices of the graph, there is a cycle of length k containing s. A cycle of this length can be obtained by successively removing triangles connected to the rest of the graph by a single edge, the removed vertex not being s, until the outer face of the remaining graph is a cycle of length k.

2.7. Non-pythagorism and arrangement of lines

Bachelard used to say that modern philosophy is a "philosophy of no", the negation being equivalent to a generalization, as in the expression "non-euclidean geometry", which is, in fact, a pan-geometry, including the euclidean geometry as a particular case. In the course of history, we always first discover particular cases. But the essence of science is to find generalizations amplifying the first discoveries, which transfigure them, sublimate them and sometimes make them unrecognizable.

As we said, the ancient Pythagoreans were fascinated by the supposed harmony of the world, which they tried to express by means of integers and their quotients – in other words, rational numbers. The emblematic figure of their sect, in which they summed up these remarks, was the particular figure of pentagram, probably because the pentagram contains many relations whose values are the golden ratio (see Parrochia 2018), as well as because there exists a duality between the regular pentagon and a star with five branches[8]. When we swap the points and lines of the first, we are inevitably led to the second (see Figure 2.27(a)).

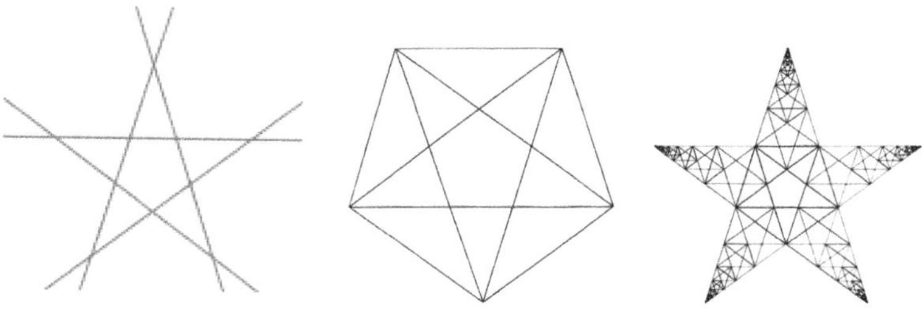

(a) (b)

Figure 2.27. *Dual arrangement and fractal pentagon (source: (b) Wikimedia commons: https://commons.wikimedia.org/wiki/File: Embedded_pentagrams_%28thick,_transparent%29.png)*

The most remarkable fact is not that a star can be embedded in the regular pentagon, but that this can be repeated (Figure 2.27(b)). You can inscribe a new star in the central pentagon delimited by the first, and so on (see Williams 1979). We would say today that we got a fractal structure. What the Pythagoreans probably saw was that the harmony was replicated to infinity and that this figure concentrated in itself the maximum possible proportionality in the world.

True or not, this particular situation shows that the arrangement of lines, i.e. configurations in plane or space, sometimes has interesting properties which can be philosophically exploited.

Roughly speaking, in geometry, an arrangement of lines is the partition of the plane formed by a collection of lines. Bounds on the complexity of arrangements

8 The figure of the pentagram was later taken up by occultism. In the thought of occultists, the passage from the pentagon (polygon with five vertices) to the pentagram (star with five branches) symbolizes the passage from the profane stage to the initiate stage. But whereas the Pythagorean figure was rather positive and – via the ratio of the sides of the isosceles triangles, equal to the golden ratio – a symbol of harmony, the pentagram of some occultists, especially when inverted, is rather a figure of disorder and chaos. This is how the figure of the pentagram was recently used by Michel Houellebecq, in his novel *Anéantir*, to designate the five places of attacks planned by an anti-tech terrorist group. In this case, indeed, it is not the straight pentagram, but the occultist inverted pentagram (i.e. the point down), which is engraved on the forehead of the Baphomet, a sort of figure of the devil. And as Houellebecq says, in agreement with many occultists, the passage from the right pentagram to the inverted pentagram symbolizes "the victory of matter over spirit, of chaos over order, and more generally of forces of evil over the forces of good" (see Houellebecq 2022, p. 569).

have been studied in discrete geometry, and computational geometries have found algorithms for the efficient construction of arrangements.

The formal definition is as follows:

DEFINITION 2.5.– *For any set A of lines in the Euclidean plane, we can define an equivalence relation on the points of the plane according to which two points p and q are equivalent if, for every line I of A, either p and q are both on I or both belong to the same open half-plane bounded by I. When A is finite or locally finite, the equivalence classes of this relation are of three types:*

1) the interiors of bounded or unbounded convex polygons (the cells of the arrangement), the connected components of the subset of the plane not contained in any of the lines of A;

2) open line segments and open infinite rays (the edges of the arrangement), the connected components of the points of a single line that do not belong to any other lines of A; and

3) single points (the vertices of the arrangement), the intersections of two or more lines of A.

These three types of objects link together to form a cell complex covering the plane. Two arrangements are said to be isomorphic or combinatorially equivalent if there is a one-to-one adjacency-preserving correspondence between the objects in their associated cell complexes.

A particular notation allows us to design all types of configuration in plane or space.

A configuration in the plane is denoted by $(p_\gamma\ \ell_\pi)$, where p is the number of points, ℓ is the number of lines, γ is the number of lines per point and π is the number of points per line. These numbers necessarily satisfy the equation:

$$p\gamma = \ell\pi,$$

as this product is the number of point-line incidences (flags).

Configurations with the same symbol, say $(p_\gamma\ \ell_\pi)$, do not need to be isomorphic as incidence structures. For instance, there exist three different $(9_3\ 9_3)$ configurations: the Pappus configuration and two less notable configurations.

In some configurations, $p = \ell$ and consequently, $\gamma = \pi$. These are called symmetric or balanced (Grünbaum 2009) configurations and the notation is often condensed to avoid repetition. For example, $(9_3 9_3)$ abbreviates to (9_3).

REMARK 2.1.– *The Fano plane (Figure 2.28, left), which exists as an abstract incidence geometry*[9]*, but cannot be constructed in the Euclidean plane, is a projective construction used by the French philosopher Dominique Dubarle to formalize Hegelian logic: this projective finite geometry is used for inscribing the infinite concept in finite determinations (see Chapter 4). But there are a lot of other similar schemes philosophers could use in the future. For example, a generalization would be the Hesse configuration of nine inflection points of a cubic curve in the complex projective plane and the 12 lines determined by pairs of these points (Figure 2.28, right). This configuration shares with the Fano plane (Figure 2.28, left) the property that it contains every line through its points. Configurations with this property are known as the Sylvester–Gallai configurations due to the Sylvester–Gallai theorem that shows that they cannot be given real-number coordinates (Kelly 1986).*

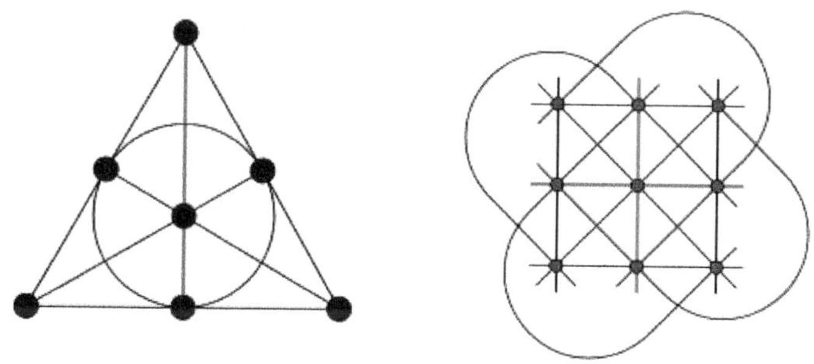

Figure 2.28. *The Fano plane and the Hesse configuration (source: left: Wikipedia article: Fano plane; right: Wikipedia article: Hesse configuration)*

Among others, we can also cite the Pappus configuration, known by Descartes, the Desargues configuration, the Schläfli double six, formed by 12 of the 27 lines on a cubic surface, the Cremona–Richmond configuration, formed by the 15 lines complementary to a double six and their 15 tangent planes, the Reye configuration, the Kummer configuration, the Gray configuration, the Klein configuration, etc.

These types of configuration were studied by Jakob Steiner. In combinatorial mathematics, we can today evaluate their complexity and study their structures via some useful tools like the Levi graph.

9 Dembowski observes that the expression "Fano plane" comes from Gleason, who imposed it despite the fact that the condition known as the Axiom of Fano (1896) forbids the occurrence of quadrangles with collinear diagonal points. For this reason, some mathematicians like Zaddach (1956–1957) call the same planes "anti-Fano planes" (see Dembowski 1968, p. 168).

2.7.1. *Levi graphs of line arrangements*

DEFINITION 2.6.– *A Levi graph or incidence graph is a bipartite graph associated with an incidence structure. From a collection of points and lines in an incidence geometry or a projective configuration, we form a graph with one vertex per point, one vertex per line and an edge for every incidence between a point and a line. They are named after Levi (1942), who wrote about them in 1942.*

REMARK 2.2.– *Pappus's hexagon theorem states that every two triples of collinear points ABC and abc (none of which lie on the intersection of the two lines) can be completed to form a Pappus configuration, by adding six lines and their three intersection points. According to Pappus' theorem, the resulting system of nine points and eight lines always has a ninth line containing the three intersection points and called the Pappus line (see Figure 2.29, left). Similarly, the Desargues configuration is a configuration of 10 points and 10 lines, with three points per line and three lines per point. It is named after Girard Desargues, and closely related to Desargues' theorem, which proves the existence of the configuration (see Figure 2.29, right).*

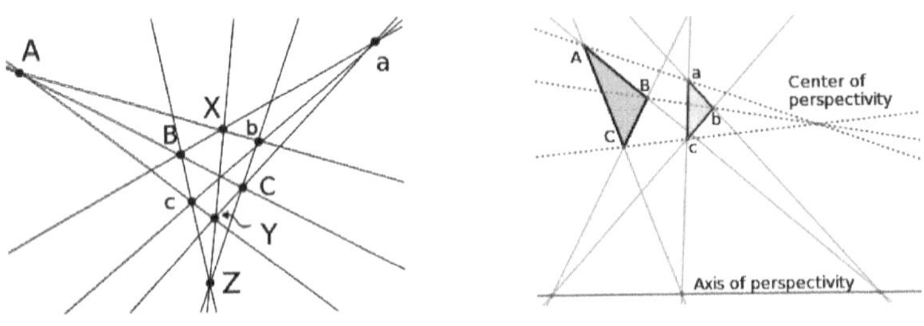

Figure 2.29. *Pappus and Desargues configurations (source: left: Wikipedia article: Pappus configuration; right: Wikipedia article: Desargues configuration)*

The first leads to the Levi graph of the Pappus configuration, also known as the Pappus graph. It is a bipartite symmetric cubic graph with 18 vertices and 27 edges (see Figure 2.30, left). The second leads to the Desargues graph, a distance-transitive cubic graph with 20 vertices and 30 edges (see Figure 2.30, right).

Another presentation of Desargue's configuration as two mutually inscribed pentagons (a first generalization of the Pythagorean graph) is shown in Figure 2.31, left). But there also exist other $(10_3 10_3)$ configurations (i.e. sets of points and lines in the Euclidean plane with three lines per point and three points per line) that are not incidence-isomorphic to the Desargues configuration. One of these is shown in Figure 2.31, right. In all of these configurations, each point has three other points that are not collinear with it. But in the Desargues configuration, these three points are always collinear with each other (if the chosen point is the center of perspectivity, then the three points form the axis of perspectivity), while in the other configuration shown in the illustration, these three points form a triangle of three lines. As with the Desargues configuration, the other depicted configuration can be viewed as a pair of mutually inscribed pentagons, a more complex construction than the Pythagorean one. Of course, the philosophical interpretation of this last configuration remains to be found.

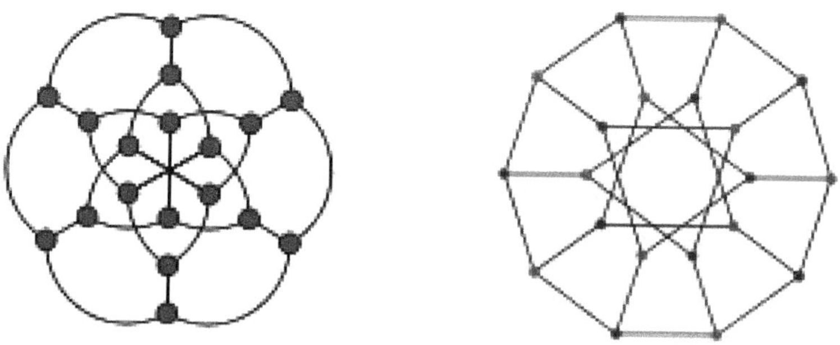

Figure 2.30. *Pappus' and Desargues' graphs (source: left: Wikipedia article: Pappus graph; right: Wikipedia article: Desargues graph)*

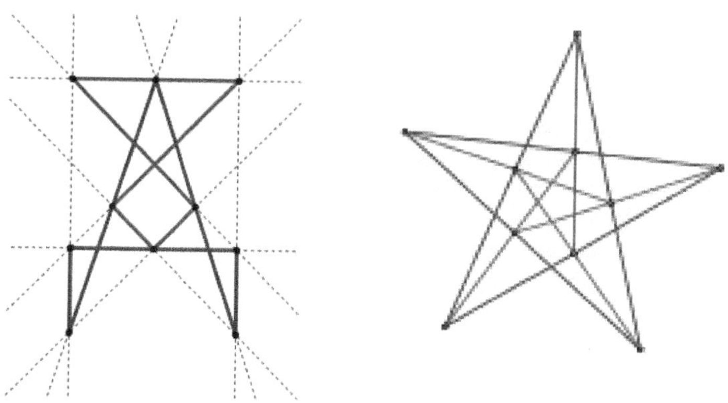

Figure 2.31. *Desargues and non-Desargues configurations (source: Wikipedia article: Desargues configuration)*

Many other Levi graphs exist, some of them showing remarkable schemes. For example, the Heawood graph (see Figure 2.32, left) is the Levi graph of the Fano plane. It is also known as the (3,6)-cage, and is three-regular with 14 vertices. The Gray graph (see Figure 2.32, right) is the Levi graph of a configuration that can be realized in $\mathbb{R}^3$ as a 3×3 grid of 27 points and the 27 orthogonal lines through them.

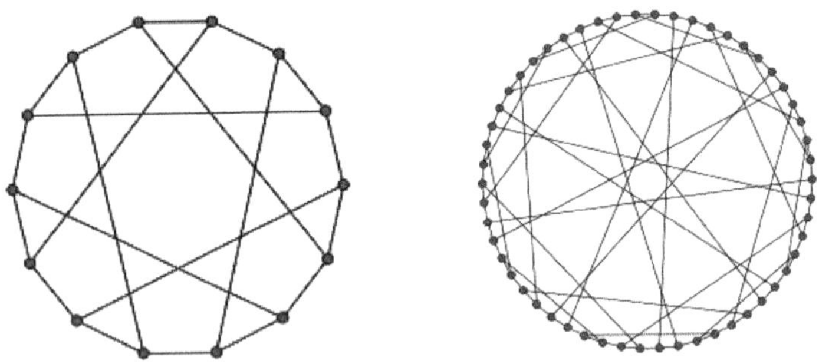

Figure 2.32. *The Heawood and Gray graphs (source: left: Wikipedia article: Heawood graph; right: Wikipedia article: Gray graph)*

Similarly, the Möbius–Kantor graph (Figure 2.33, left) is the Levi graph of the Möbius–Kantor configuration (Figure 2.33, right), a system of eight points and eight lines that cannot be realized by straight lines in the Euclidean plane. It is three-regular with 16 vertices.

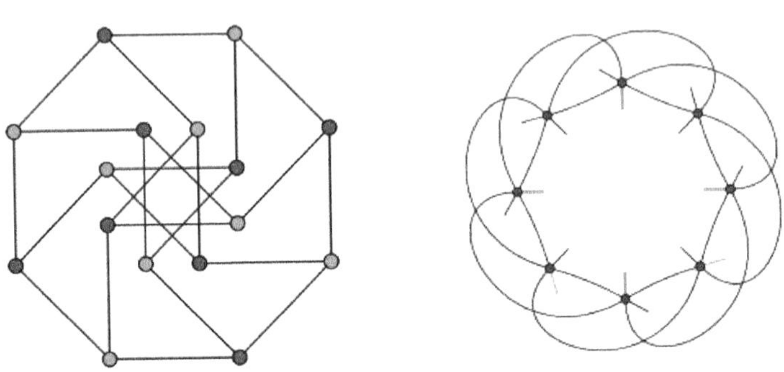

Figure 2.33. *The Möbius–Kantor graph and the corresponding configuration (source: left: Wikipedia article: Möbius-Kantor graph; right: Wikipedia article: Möbius-Kantor configuration)*

We may also mention the 12-vertex crown graph (see Figure 2.34, left), which is the intersection graph of the lines of the Schläffli double six (see Figure 2.34, right)[10].

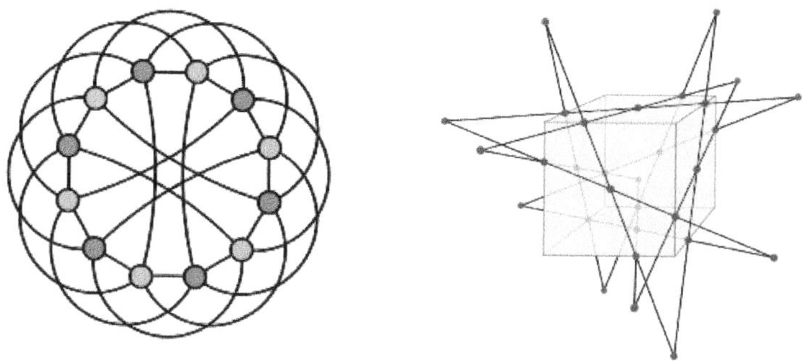

Figure 2.34. *The 12-vertex crown graph and the Schläfli double six (source: left: Wikimedia commons: https://commons.wikimedia.org/wiki/File:Crown_graph_12.svg; right: Wikipedia article: Schläfli double six)*

2.7.2. *Line arrangements of curve lines*

A line may be a curve. So, there also exists the arrangement or pseudo-arrangement of curves, leading equally to some kinds of Levi graphs. They generally refer to non-euclidean geometries.

So is, for example, the Tutte eight-cage (see Figure 2.35, left) which is the Levi graph of the Cremona–Richmond configuration (see Figure 2.35, right). It is also known as the (3,8)-cage, and is three-regular with 30 vertices.

REMARK 2.3.– *The Schläfli double six, the Cremona–Richmond configuration and the cubic surfaces (which are del Pezzo or Fano surfaces) are connected. A generic cubic surface contains 27 lines, among which 36 Schläfli double six configurations can be found. The set of 15 lines complementary to a*

10 Schläfli (1858, 1863) found cubic surfaces containing sets of 15 real lines (complementary to a Schläfli double six in the set of all 27 lines on a cubic) and 15 tangent planes, with three lines in each plane and three planes through each line. Intersecting these lines and planes by another plane results in a 153,153 configuration. The specific incidence pattern of Schläfli's lines and planes was later published by Cremona (1868). The observation that the resulting configuration contains no triangles was made by Martinetti (1886), and the same configuration also appears in the work of Richmond (1900). Visconti (1916) found a description of the configuration as a self-inscribed polygon.

double six, together with the 15 tangent planes through triples of these lines, has the incidence pattern of another configuration, the Cremona–Richmond configuration.

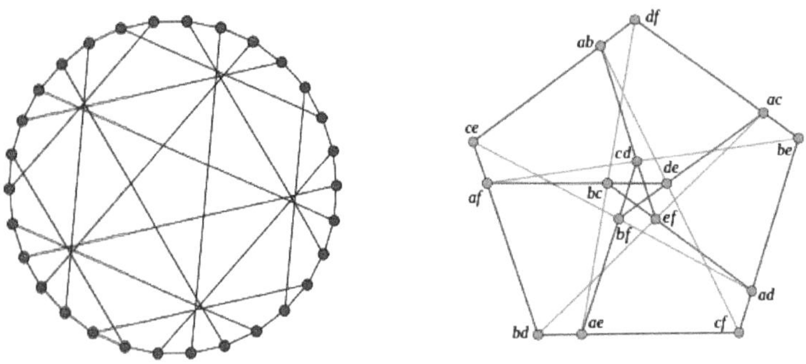

Figure 2.35. *The Tutte eight-cage and the Cremona–Richmond configuration (source: left: Wikipedia article: Tutte-Coxeter graph; right: Wikipedia article: Cremona–Richmond configuration)*

The intersection graph of the 12 lines of the double six configuration is a 12-vertex crown graph, a bipartite graph in which each vertex is adjacent to five out of the six vertices of the opposite color. The Levi graph of the double six may be obtained by replacing each edge of the crown graph by a two-edge path. The intersection graph of the entire set of 27 lines on a cubic surface is the complement of the Schläfli graph.

Clebsch gave a model of a cubic surface, called the Clebsch diagonal surface, where all 27 lines are defined over the field $\mathbb{Q}[\sqrt{5}]$*, and in particular are all real (see Figure 2.36).*

The 27 lines can also be identified with some objects arising in the representation theory. In particular, these 27 lines can be identified with 27 vectors in the dual of the E_6 *lattice, so their configuration is acted on by the Weyl group of* E_6*. In particular, they form a basis of the 27-dimensional fundamental representation of the group* E_6*.*

These 27 lines can be identified with the 27 possible charges of M-theory on a six-dimensional torus (six momenta; 15 membranes; six five membranes) and the group E_6 *then naturally acts as the U-duality group. This map between del Pezzo surfaces and M-theory on tori is known as "mysterious duality".*

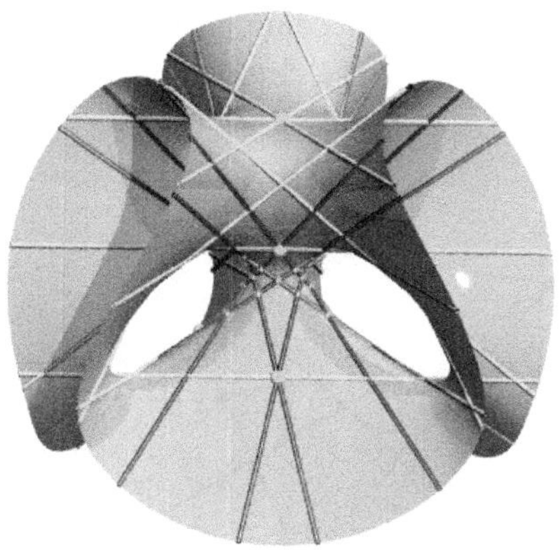

Figure 2.36. *The Clebsch surface (source: https://mathcurve.com/surfaces.gb/clebsch/clebsch.shtml)*

We may then introduce the following definition:

DEFINITION 2.7.– *A pseudoline arrangement is a family of curves that share similar topological properties with a line arrangement.*

These can be defined most simply in the projective plane as simple closed curves, any two of which meet in a single crossing point. A pseudoline arrangement is said to be stretchable if it is combinatorially equivalent to a line arrangement. It is complete for the existential theory of the reals to distinguish stretchable arrangements from non-stretchable ones. Every arrangement of finitely many pseudolines can be extended so that they become lines in a "spread", a type of non-Euclidean incidence geometry in which every two points of a topological plane are connected by a unique line (as in the Euclidean plane), but in which other axioms of Euclidean geometry may not apply. We may give the following example (see Figure 2.37).

2.7.3. *Hyperbolic graphs*

Another type of non-Euclidean geometry is the hyperbolic plane, and arrangements of hyperbolic lines in this geometry have also been studied. Any finite set of lines in the Euclidean plane has a combinatorially equivalent arrangement in the hyperbolic plane (e.g. by enclosing the vertices of the arrangement in a large circle and interpreting the interior of the circle as a Klein model of the hyperbolic plane). However, in hyperbolic line arrangements, lines may avoid crossing each other without being parallel. The intersection graph of the lines in a hyperbolic arrangement is a circle graph.

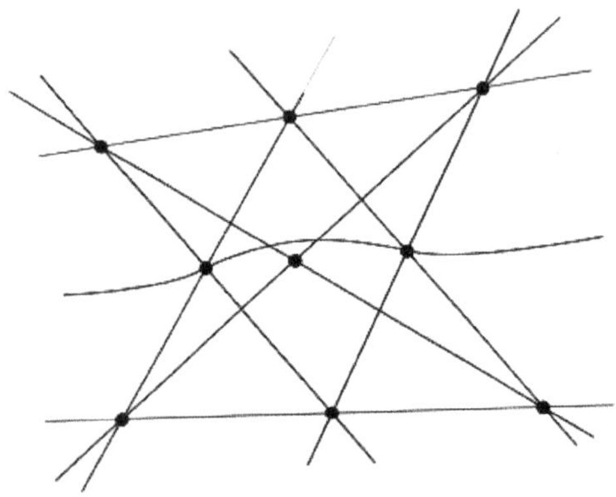

Figure 2.37. *Pseudo-arrangement of nine pseudolines (source: Wikipedia article: Arrangement of lines)*

DEFINITION 2.8.– *In graph theory, a circle graph is the intersection graph of a set of chords of a circle. That is, it is an undirected graph whose vertices can be associated with chords of a circle such that two vertices are adjacent if and only if the corresponding chords cross each other. See an example of that in Figure 2.38, with the associated graph.*

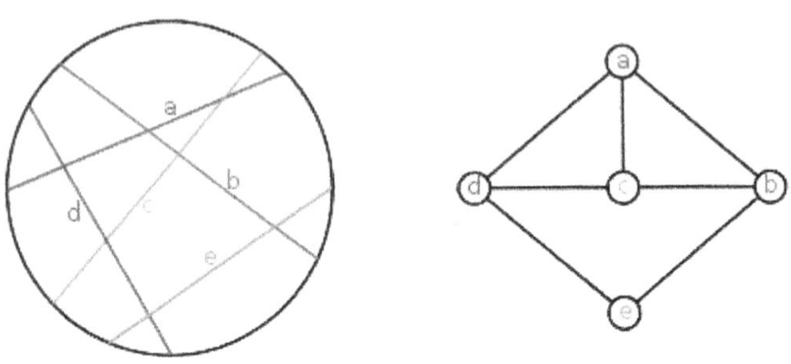

Figure 2.38. *A circle with five chords and the corresponding circle graph (source: Wikipedia article: Circle graph)*

The corresponding concept to hyperbolic line arrangements for pseudolines is a weak pseudoline arrangement, a family of curves with the same topological properties as lines, such that any two curves in the family either meet in a single crossing point or have no intersection (see Figure 2.39).

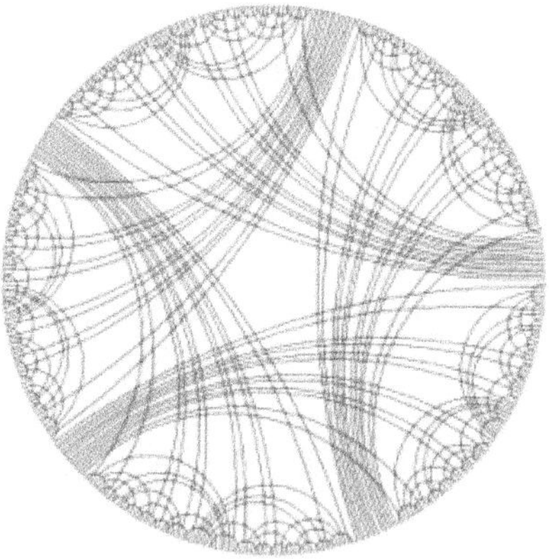

Figure 2.39. *Hyperbolic arrangement (source: Wikipedia article: Arrangement of lines (color modified))*

Avoiding intersections or some particular figure like triangles must be a requirement for a philosophical architecture. This is possible with some particular kinds of line arrangement.

REMARK 2.4.– *In Figure 2.37, we may remark, in this triangle-free circle graph (see Ageev 1996), a family of curvy pentagons which recall the pythagorean one and constitute a non-euclidean (in fact hyperbolic) generalization of it. This can be considered as an ultimate avatar of the Pythagorean pentagram.*

This curved pentagon is a distant generalization of the Pythagorean pentagon at the origin of philosophy. But, to our knowledge, no philosopher, until now, has made use of it. All that can be said is that this kind of structure, which generalizes that of a mathematical tree, could be proposed to model the structure of information in the human brain.

Recall that a mathematical tree is a connected graph without cycles (i.e. without loops), defined by the following conditions:

1) no point (respectively, subset of points) of a tree is isolated. The loss of this property of "connectedness" makes it possible to define the mathematical concept of a "forest": a forest is a graph without cycle, but which is not connected;

2) any point of a tree is connected to another by an edge and only one. It is not possible, starting from a point and following a series of edges in the same direction, to return to the point from which we started;

3) the number of vertices in a tree is equal to the number of edges minus 1;

4) a tree with root is a "rooted tree". But this property is only optional. Unlike real trees, a mathematical tree does not necessarily have a root, and when it does, it is not necessarily placed at either end.

As we know, these properties make tree structures well suited to the representation of knowledge. The mathematical tree is particularly useful for dealing with problems of classification and scheduling of tasks (PERT graphs), or even, for solving questions of a shorter path (tree of minimum weight) or of a longer path (tree of maximum weight), i.e. optimization problems in operations research. Algorithms and linear programs are then associated with the automatic search for solutions.

Recently, however, this structure may have appeared insufficient to characterize cerebral structuring, in particular that linked to "long memory". But, as the French mathematician Alain Connes (Fields Medal 1982) reminded us, the fundamental properties of mathematical trees are not specific and more general structures also have them. Thus, the property of the "shortest path" is found in the notion of "hyperbolic simplicial complex", even though the uniqueness of the path from one point to another is no longer necessary. This structure thus becomes a generalization of the notion of tree. Topologically, it has the same properties as a tree without having the drawbacks.

We can explain these remarks precisely as follows.

An abstract simplicial complex is the data of a set V (whose elements are called vertices) and of a set Σ of non-empty finite parts (the faces) of V, which is stable by non-empty sub-parts. In other words, any non-empty part of a face is also a face. It is generally required that any vertex belongs to a non-zero finite number of faces.

To define now a simplicial hyperbolic complex, we can of course use combinatorial criteria, but, following Connes (see Changeux and Connes 1989, pp. 173–188), it is more meaningful to start from geometry and the notion of geodesic. If we consider the Poincaré model of non-Euclidean geometry (the interior of a disc in the plane), the geodesics (or shortest path line) are the arcs of circles perpendicular to the edge of the disc.

If we take such a geodesic, and a point P outside it, it is easy to construct an infinity of other geodesics passing through P and not meeting the first one. This geometry therefore does not verify Euclid's axiom of parallels. In this model, the angle between two geodesics is the angle between the corresponding circles. We also verify that the sum of the angles of a triangle is always less than 180°, a property of a space of negative curvature. We can also specify how to measure the distance between two points in this geometry. The shortest path between two points A and B is precisely the geodesic, i.e. the portion of the circle which passes through the two points perpendicular to the edge of the disc. This geometry has a property which brings it closer to the tree and which Euclidean geometry does not have at all (see Figure 2.40).

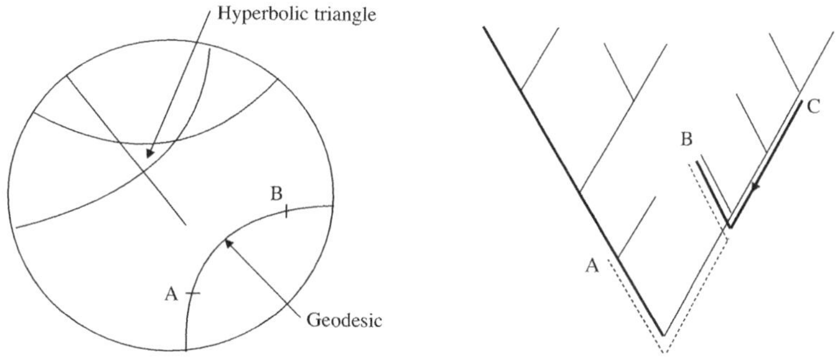

Figure 2.40. *Poincaré hyperbolic space (a) and a tree (b)*

This is precisely hyperbolicity. A simple way of expressing it is to say that if BC is a segment and A is a point outside this segment, then when we want to move from A to B, we lose very little, and in any case not more than a quantity fixed once and for all, with respect to the optimal displacement given by the geodesic, by replacing it with the shortest path joining A to the segment BC, followed by a displacement along BC. This property, false in Euclidean geometry, is true for trees, and also true for the Poincaré hyperbolic space. Moreover, it continues to be true for the universal covering of a very large number of simplicial complexes which are precisely the hyperbolic simplicial complexes.

We deduce, for example, that a database based, not on trees but on simplicial hyperbolic structures, would allow greater research flexibility in the search for information. Modern philosophical systems could therefore adopt this structure, the graph of which in Figure 2.40 is a particular example.

3

Order and Its Philosophical Use

In this chapter, essentially intended to recall some basic notions, we will discuss some very general problems concerning the mathematical notion of order, which will guide our study and will constitute, we admit, one of its main motivations. But we first want to explain why this notion of "order" is essential for the philosopher, and why philosophers should be interested in its formalization.

The (informal) notion of order is very old in philosophy. The word "order" (Latin *ordo, -dinis*) is very polysemic in west languages, the meanings ranging from the injunction (the order to which, for example, a soldier obeys), and going to the designation of an assembly guaranteeing the rights of a given community (the order of lawyers, doctors, notaries, etc.).

The meaning to which we will attach (because it is the core of all the others) is that of organization, arrangement, disposition or distribution presenting a certain regularity – whose succession is only a case in particular. Whether this provision is objective or, on the contrary, is only perceived as satisfactory for the mind, is certainly a philosophical problem, but we will not deal with this here.

Philosophy has always been defined as an attempt to "put the whole of thought in order": this is even the essence of the philosophical project according to philosopher and historian Bernard Bourgeois, a specialist in German philosophy and, more specifically, Hegelian thought (see Bourgeois 1970).

Throughout history, this project has declined in various ways according to the different contents of this thought. Here are some examples:

1) ordering the soul and its affections (Spinoza, Freud);

2) ordering the city and its institutions (Plato, Hobbes, Rousseau, etc.);

3) putting in order the whole "world" or the knowledge we get from it (Descartes, Kant, etc.);

4) ordering history (theories of progress: Hegel, Comte, etc.).

This notion of order is also very present in the natural sciences (establishment of series, chronologies, family trees, taxonomies, rankings, etc.).

A pre-established question may therefore arise: do we know what this notion of order is, whose existence we postulate and are these operations of ordering all these intelligible or sensitive domains always possible?

In the quest that led philosophers to implement such a project, it seems that they have made many illusions about the result, precisely because they did not perceive the difficulties related to the various notions of order that we can imagine, and which have only gradually revealed themselves to the mathematicians in history.

To go a little further, we must necessarily access to the mathematical sense of the notion of order.

3.1. The mathematical notion of order: a brief history

Many concepts relating to the mathematical notion of order appear at the end of the 19th century.

It is, for example, the notion of "successor of a number" in the axiomatic of Peano, or the notion of "dense order", almost already present in Dedekind in his characterization of the set $\mathbb{Q}$ of rationals and in its construction of reals.

Similarly, during the 1880s, Georg Cantor, in his reflection on the foundations of set theory and especially on the theory of ordinals, already implicitly considered what would be called "linearly ordered sets" and "well-ordered sets" today. It will be seen in a moment that these are "total orders". But, as will be seen too, there are also so-called "partial orders", i.e. structures which tolerate the existence, in the same "ordered set", of incomparable elements.

It does not seem, however, that mathematicians have really been interested in the notion of "partial order" before the beginning of the 20th century, although we may find some traces of it in the works of Charles Sanders Peirce in 1885 (see

Peirce 1885) and Schröder (see Schröder 2000) in the years 1890–1905 on the algebra of logic.

Among the first works on partial orders in the first half of the 20th century are the famous article by MacNeille on the partially ordered sets (see MacNeille 1937) and Birkhoff's classic work on the concept of lattice (see Birkhoff 1940). A second revised edition was published in 1948 and a third in 1967.

Let us also mention the work of Duro (George) Kurepa on this particular class of partial orders that are trees, and especially infinite trees (Kurepa 1935).

The theory of orders will then be developed in a substantial way by Felix Hausdorff (1949), while Dushnik and Miller (1941) had already elaborated a reflection on the dimension of ordered sets.

At the end of the 1950s came the first exclusive contribution to a general theory of ordered sets, that of the Spanish mathematician Norberto Cuesta Dutari (1959).

But it was not until the 1980s that important books appeared on the subject, such as Marcel Erné's *Einführung in die Ordnungstheorie* (1982), Roland Fraïssé's *Theory of Relation* (1986), Peter C. Fishburn's *Interval orders and interval graphs* (1985) and a scientific journal (ORDER) entirely devoted to the question of orders.

From the 1990s, order theory as such has its place in mathematics. Among others are the classical book of B. A. Davey and H. A. Priestley, *Introduction to Lattices and Order* (2012), a panoramic article by William T. Trotter in the *Handbook of Combinatorics* (1995) and, more recently (2003), a book summarizing a large part of the knowledge on the subject: *Ordered Sets* by B. W. S. Schröder (2002). Also in 2005 was published a very interesting textbook entitled *Ordered Sets* by E. Harzheim (2005).

The work of previous mathematicians has resulted in a terminology that can be summarized as follows.

Let us recall first that a relation $\mathcal{R}$, defined on a set E, is:

– *reflexive* if $\forall x \in E, x\mathcal{R}x$;

– *irreflexive* if $\forall x \in E, x \not\mathcal{R} x$;

– *weakly antisymmetric* if $\forall(x, y) \in E^2$, $[(x\mathcal{R}y) \wedge (y\mathcal{R}x)] \Rightarrow (x = y)$;

– *strongly antisymmetric* if $\forall(x, y) \in E^2, (x\mathcal{R}y) \Rightarrow (y \not\mathcal{R} x)$;

– *transitive* if $\forall(x, y, z) \in E^3, [(x\mathcal{R}y) \wedge (y\mathcal{R}z)] \Rightarrow (x\mathcal{R}z)$.

Now a relation $\mathcal{R}$, defined on a set E, provides E with a structure of:

1) *preorder* (or *quasiorder*) if it is reflexive and transitive;

2) *strict order* if it is irreflexive and transitive;

3) *order* (short) if it is reflexive, antisymmetric and transitive;

4) *linear order* (or *total order*) if $\mathcal{R}$ is a relation of order, and if, moreover, for any element $a \neq b$, either $a\mathcal{R}b$ or $b\mathcal{R}a$.

When elements of the ordered set remain incomparable, the order is called a *partial order* and the corresponding set a *partial ordered set* or *poset*.

DEFINITION 3.1.– *Let $(P, \leq)$ be a poset. A subset $T \subseteq P$ is called a chain (resp. an antichain), if every two different elements of T are comparable (resp. incomparable). The length of a finite chain is the same as its cardinality. But, of course, chains may be infinite.*

3.2. The idea of "well-ordering"

DEFINITION 3.2 (WELL-ORDERED SET).– *We say that a linearly ordered set E is well-ordered or is a well-ordering if every non-empty subset of E has a smaller element.*

This amounts to saying that there is no infinite downward sequence of different elements on this set E. In other words, there is not, on E, a descendant sequence $S = a_0, a_1, ..., a_n, ...| \; n < \omega$ such that $a_0 > a_1 > a_2 > ...$: it is still said that E is well founded.

The problem that immediately arises, just by limiting itself to this notion, is whether or not the set E, "well-ordered" in the sense of the relation of "well-ordering" defined above, has a smaller element. The answer of mathematicians is, apparently, positive, in the sense that we have, on this subject, a theorem, proved by Zermelo at the beginning of the 20th century.

THEOREM 3.1 (ZERMELO, 1904).– *Every set can be well-ordered.*

PROOF.– Let A be a set. To define an order on A, it is necessary to construct a transfinite sequence: $\langle a_\alpha : a < \theta \rangle$ that we will put in bijection with the elements of A and which will enumerate them. This is done by induction, using what is called a "choice function" f associating with an ordinal integer – the smallest possible – any non-empty subset of A. So, the sequence in question lists A. □

There is nevertheless a problem: this proof presupposes, in fact, that we can construct one set (the set that enumerates) by choosing elements in some others (the collection of ordinal numbers). In other words, Zermelo's theorem presupposes something indemonstrable in set theory, and that has been established, for this reason, as an axiom: the axiom of choice.

We must even recognize that there is a total equivalence (which, moreover, is proved) between "well-ordering", on the one hand, and "make a choice", on the other hand.

So, we must face an embarrassing consequence: every set can be well-ordered if, to obtain its elements, we have to make choices whose set is itself well-ordered.

This amounts to stating a kind of tautology, namely, that "every set can be well-ordered if and only if it can be well-ordered".

It is clear that such a statement poses obvious limits to the relevance of establishing a "well-ordering" on anything infinite: so we say goodbye to the reformers of the soul, the theoreticians of progress, the defenders of a natural order in the world, etc. Nothing really consistent may justify their hierarchies, their serialization or their devices of causal type: nothing in any case, which can be mathematically founded, especially if they deal with infinite sequences. Let us see, on this subject, a very famous example.

Spinoza's claim: In his *Ethics* (Prop. X, Book 5), Spinoza declares:

> As long as we are not dominated by affections which are contrary to our nature, we have the power to order and chain the affections of the body according to a valid order for the understanding.

Unfortunately, no! In Spinoza's philosophy, each soul, which is a finite mode, is inserted in the infinite mode of Thought, an effect or consequence of the corresponding attribute of Thought, one of the infinite attributes of God. If each soul had this possibility to be well-ordered, the infinite set of finite modes would also be well-ordered, which, according to Zermelo's set theory, supposes the axiom of choice. So nothing proves that this is more than a postulate[1].

We could think of escaping these disastrous consequences by weakening the demand.

A first way of doing this would be to transform the axiom of choice into an axiom of the countable choice (only, at that moment, any countable family of non-empty sets would have a function of choice). But this limitation does not change anything.

A second way would be to replace the axiom of choice with Zorn's lemma:

> If every non-empty chain of a partially ordered set S has an upper bound in S, then S has a larger element.

But this lemma is in fact totally equivalent to the axiom of choice: the possibility of establishing an order on every set is, in fact, postulated, but not, strictly speaking, demonstrated.

Alternative theories do not go farther.

Let us forget ZF and consider, for example, the Gödel constructible set theory which posits a hierarchy L of sets such that:

1) $L_0 = \emptyset$;

2) $L_\alpha = U_{\beta<\alpha} = L_\beta$ if α is a limit ordinal;

3) $L = U_{\alpha \in Ord L_\alpha}$.

---

1 With the notion of order, in the mathematical sense of the term, we can specify here the Spinozist structure of nature. The correspondence between the plane of ideas of thought and that of extension forms, as we have seen, is a bipartite graph. But, if we define a total order structure (chain) on these two infinite modes, then, we get what is called a *bichain*. A result of Pouzet (1978) then shows that there exists a countable bichain named *universal bichain* in which all the countable bichains are embedded (by order-preserving maps). This bichain is easy to describe: it is, for example, the bichain obtained by calculating the ordinal product of the bichain $Q_{\alpha,2}$ by the bichain Q of the rationals. The bichain $Q_{\alpha,2}$, with α irrational, being formed of the pairs (x, i) with x rational and i equal to 0 or 1, and ordered as follows: $(x, i) \leq (y, j)$ when either $i = j$ and $x \leq y$ in Q, or $i \neq j$ and $x + \alpha \leq y$ in $\mathbb{R}$. In this context, the Spinozist structure of nature is an n-chain. But we ignore if we can find, for the n-chains, the result obtained on the 2-chains. Moreover, Spinoza favoring continuity, it is to be feared that his natural n-chain is not countable, but has, in fact, the power of the continuum.

Certainly, Gödel's model proves the existence of a well-ordering of the L class, and thus derives the axiom of choice, but it is at the cost of the surreptitious introduction of a hierarchy $V = L$. It is very peculiar and, here again, presupposes what is to be proved.

In fact, set theorists are close to think that $V = L$ has a very limited interest.

The hierarchy $V = L$ consists of admitting as subsets of a set only those that can be clearly defined. One consequence is that Gödel's hierarchy of building complexes is beautiful, perfect, and an absolute order, but basically too good to be true, so no one believes it.

There is no serious mathematician today who admits the axiom of constructibility. And why? Because this axiom freezes the theory of sets and makes it practically uninteresting. It is much more interesting to find non-constructible sets, as Cohen proved it can be done with the so-called "generic" sets, that escape this huge total Gödelian order which consists of establishing a kind of military structure on everything.

In conclusion, it appears that mathematics is therefore unable to demonstrate the possibility of putting order in any (infinite) set, which does not mean, of course, that the enterprise is impossible in the finite. Mendeleev, for example, ordered the chemical elements linearly (the atomic numbers instantiating a finite sequence of ordinals). But this finite "well-ordering" is only the equivalent of a good choice, maybe a bet on the continuity of nature, which here has succeeded. Nothing says that this must always be the case, and so, our desire to establish some ordering (especially a "well-ordering") in the world or, say, in the infinite sets that represent it, is a simple possibility and cannot be, in any way, an absolute necessity.

3.3. Quasiorders (or preorders)

We may wish to escape from the uncertainty thrown on the philosophical projects of ordering the world because of the impossibility of proving the existence of a "well-ordering" on a set, by weakening the initial project. Now, we do not want things to be necessarily linearly ordered anymore. We only want them to be able to enter what are called, in mathematics, quasiorders or preorders.

Remember that a set is a quasiorder if there exists between its elements only a reflexive and transitive relation. There is for the quasiorders the same type of weakening as for the orders.

DEFINITION 3.3.– *A set E is said to be well-quasiordered or a well-quasiorder, if:*

1) as in the case of well-orders, there is no infinite descending chain of distinct elements on E (in other words, every non-empty subset of E has a smaller element);

2) moreover, there must not be on E infinite sets of antichains (we call an antichain a set of pairwise incomparable elements).

One of the first results on well-quasiorders was Higman's theorem, proven in 1952:

THEOREM 3.2 (HIGMAN).– *The class of finite linear orders (chains) indexed by a finite set is a well-quasiorder for inclusion.*

The objects of the class are finite chains whose elements are indexed by a finite set, and the inclusion preserves order and indexing by injection. For example, the chain *aabcc* is included in this sense in *abcabcabc* but not in *cbacbacba*.

This result was extended by Kruskal in 1960 to the class of finite trees, then by Nash-Williams (1965) to that of infinite trees, finally to that of finite graphs by Robertson and Seymour (1986–1996).

The passage to infinity, alas, poses a problem in the case of quasiorders, because we know, since a result of Rado (1954), that we can build an infinite set of incomparable countable subsets of a well-quasiorder. This is Rado's famous counterexample of partially ordered sets of vectors.

This obviously limits the interest of the notion of "well-quasiorder" when we want to explore the infinite because Rado shows in fact that a quasiorder on an infinite set is not incompatible with a kind of internal disorder of its subsets.

To overcome the limits imposed by the counter-example of Rado, Nash-Williams (1968) introduced two fundamental tools in the theory of quasiorders:

1) the *minimal bad sequence* which has, in particular, a methodological role and makes it possible to shorten the evidence concerning the existence of well-quasiorders;

2) the notion of *better-quasiordering* which is a reinforcement of the notion of "well-quasiordering" and which provides an effective tool for addressing countable structures.

Examples of well-quasiorderings: In this context, Laver (1971) proved that the class of infinite countable chains is "better-quasiordered", and thus "well-quasiordered" with respect to inclusion. Later, Corominas (1982) extended the Kruskal result for finite trees to countable trees. In other words: the class of countless infinite trees, indexed by a better-quasiordering, is better-quasiordered with respect to inclusion.

Maurice Pouzet (1985) then posed the problem of another class of orders, the N-free orders or orders formed with parallel series (a class which contains the class of trees).

Pouzet succeeded in proving that the class of N-free finite orders is well-quasiordered and then conjectured that of infinite countable N-free orders should be better-quasiordered.

This last conjecture was proven by his student Stéphane Thomassé in 1995.

We thus obtain significant results in the infinite with countable better-quasiorders.

Generally, if you can define a better-quasiordering on a set, then you can also get a well-quasiorder on it. But the definition of a better-quasiordering, reinforcing the notion of "well-quasiorder", presupposes in turn the axiom of choice.

In conclusion, mathematics, again, gives us no assurance that the sensible can be, in general, better-quasiordered, and therefore, well-quasiordered.

When we obtain such a result, we must know that it is only a rewrite of what we have posited in axiom with the possibility of choosing, i.e. of ordering the elements of a set.

3.4. Partial orders

As we know, a relation $\mathcal{R}$, defined on a set E, is an *order relation* if it is reflexive, antisymmetric and transitive. Now, if some pairs of elements of E are incomparable, then $\mathcal{R}$ equips E with a structure of *partial order*. The philosophical meaning of this notion is clear: to reduce ourselves to a "partial order" is in fact to tolerate some kind of non-order in order.

Formally speaking, a (finite) partially ordered set, also known as a (finite) poset (see Stanley 2004, p. 97), is a finite set P together with a binary relation $\leq_P$, which is often denoted $\leq$ if there is no confusion, such that:

1) for all $x \in P$, $x \leq x$ (reflexivity);

2) if $x \leq y$ and $y \leq x$, then $x = y$ (antisymmetry);

3) if $x \leq y$ and $y \leq z$, then $x \leq z$ (transitivity).

We use the obvious notation $x < y$ to mean $\leq y$ and $x \neq y$. Similarly, $x \geq y$ and $x > y$ stand for $y \leq x$ and $y < x$, respectively.

As the uniquely determined minimal transitive reduction of the poset P, its Hasse diagram is the digraph $\Gamma(P)$ with P as the vertex set and there is an arc from y to x if y covers x, i.e. if $||[x, y]|| = 2$. In this case, we denote the order relation by $x \lessdot y$.

Partially ordered sets (posets) are the most general types of orders because they possess many generic properties, some of which they share with orders or quasiorders.

Examples of posets:

1) the lattices (partial orders where every pair of elements have a unique least upper bound and a unique great lower bound);

2) the semilattices (partial order where every pair of elements have a unique least upper bound or a unique great lower bound, but not both);

3) trees (graphs without cycles);

4) quasiorders;

5) total orders or chains (chains of elements, chains of subsets, chains of partitions, etc.) are, of course, a fortiori, partial orders.

Problems: The problems raised by partial orders are very numerous.

For example, it is a question of whether there are single decompositions of complex orders into simpler elements that "factorize" them.

It is also about understanding the links existing between partial orders or between subsets of a partial order like their chains or their anti-chains.

Tools for studying partial orders: As a partially ordered set is a digraph, we can use, for studying its properties, some tools of graph theory. For instance, we can determine a matrix which is the Boolean representation of the corresponding Hasse diagram. We may define it in different ways.

First, following Stanley (see Stanley 1970, 1988, 2004) and more recent searchers (see Wu and Shizen 2012), we shall say that the incidence matrix $\mathfrak{m}_P$, which encodes full information about the poset P, is the indicator function of $<$ and may be described more explicitly as follows:

$$\mathfrak{m}_P = \begin{cases} 1, & \text{if } x < y; \\ 0, & \text{otherwise.} \end{cases}$$

Similarly, the covering matrix $\mathfrak{c}_P$ is defined as follows:

$$\mathfrak{c}_P = \begin{cases} 1, & \text{if } x \lessdot y; \\ 0, & \text{otherwise.} \end{cases}$$

An alternative to this representations was reported by Rhodes and Silva (2015) who have adopted the following notation:

$$\mathfrak{m}'_P = \begin{cases} 0, & \text{if } x \geq y; \\ 1, & \text{otherwise.} \end{cases}$$

Suppose you get for P the following Hasse diagram of Figure 3.1:

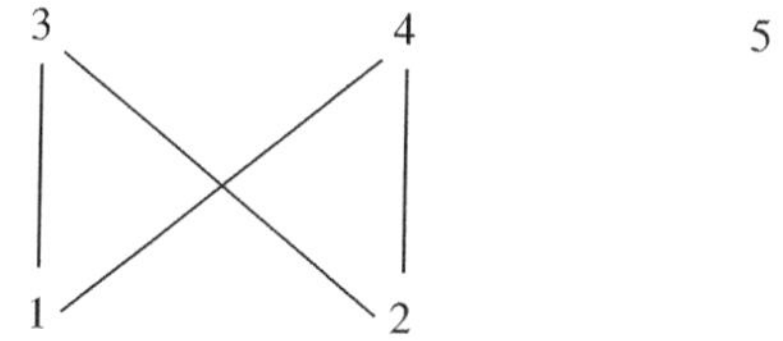

Figure 3.1. *Hasse diagram of the poset P*

Then, the incidence matrix $\mathfrak{m}'_P$ is such that:

$$\mathfrak{m}'_P = \begin{pmatrix} 0 & 1 & 1 & 1 & 1 \\ 1 & 0 & 1 & 1 & 1 \\ 0 & 0 & 0 & 1 & 1 \\ 0 & 0 & 1 & 0 & 1 \\ 1 & 1 & 1 & 1 & 0 \end{pmatrix}$$

There is therefore plenty of choice to characterize the partial orders. We will choose the most appropriate formula for the desired goals.

Now, assume that a philosophical system is such that its basic categories are organized according to some partially ordered set, we can get the incidence matrix of it and study its fundamental properties as if it were only a simple graph.

In the following, we will limit ourselves to a class of partial orders that share good properties with finite partial orders: well partial orders and trees.

3.4.1. *The notion of well partial order*

Remember that a quasiorder $(W, <)$ is said to be *well founded* if it does not have an infinite descending chain of elements, i.e. if there is not $a_0, a_1, ..., a_n, ...|n < \omega$ such that $a_0 > a_1 > a_2 > ...$

As we have already seen, if W is well founded and if, moreover, it does not have an infinite antichain, then it is a well-quasiorder (or wqo).

Now, if W is not only a quasiorder but an order, then, if W is well founded and if W does not have an infinite antichain, W is a well-quasiorder.

3.4.2. *Linear extension of a poset*

Can we linearly order a partially ordered set? Apparently, the answer is "yes". It seems sufficient to define some linear extension containing the partial order.

DEFINITION 3.4 (LINEAR EXTENSION).– *Let $(P, \leq)$ be a poset, a and b be two incomparable elements of P. Then, there exists a relation of order $\leq^*$ on P which contains $\leq$ and for which we have $a \leq^* b$. Such a relation $\leq^*$ is named linear extension of $\leq^*$.*

THEOREM 3.3 (SZPILRAJN 1930).– *Let $(P, \leq)$ be a poset. Then, there is a linear order $\leq$ on P, which contains the extension $\leq^*$ of $\leq$.*

This last theorem shows that we can linearly order any partially ordered set. But such a result is actually obtained under the axiom of choice.

3.4.3. *Well partially orderings*

THEOREM 3.4 (WOLK 1967).– *A partially ordered set* $(P, \leq)$ *is well partially ordered (wpo) if and only if any linear extension of* $\leq$ *is a well-order on* P.

THEOREM 3.5.– *A partially ordered set* $(P, \leq)$ *is well founded if and only if there is a linear extension of* $\leq$ *which is a well-order of* P.

As can be seen, the possibility of ordering or founding a partial order is based on the existence of linear extensions, i.e. again, on the axiom of choice.

THEOREM 3.6.– *If* $(P, \leq)$ *is well partially ordered, then* P *has no infinite antichain.*

THEOREM 3.7.– *Any finite partially ordered set is well partially ordered.*

The proof is obvious. As every finite set can be well-ordered, every finite well-ordered set is, a fortiori, partially ordered.

This last property shows that, on the condition of being restricted to the finite case, all sets can be well partially ordered, i.e. put in the form of a well partial order.

3.5. Trees

From the viewpoint of graph theory, a tree may be defined as a connected graph without cycle. In order theory, a tree is a partially ordered set, and generality imposes to consider infinite trees. From this, we get the following definition:

DEFINITION 3.5 (TREE).– *A partially ordered set that has a first element* x, *an infinity of incomparable pairwise neighbors above* x, *and no other element, is a tree.*

THEOREM 3.8.– *A tree is well-ordered and well founded.*

REMARK 3.1.– *Unlike a well partially ordered set, a tree can have infinite antichains.*

REMARK 3.2.– *The description of the properties of trees suggested to be interested in the cardinality of chains, antichains and levels, which did not fail to do mathematicians.*

3.5.1. *A zoo of infinite trees*

The study of infinite trees began around the middle of the 1930s, in relation with set theory and Suslin's problem dating to 1920[2].

THEOREM 3.9 (KÖNIG).– *Let T be an infinite tree, all levels of which are finite. So, there is an infinite branch in this tree.*

The immediate generalization of this theorem asked a big question:

Let T be a tree with $\aleph_1$ elements and of which all levels are countable. Does such a tree necessarily have a branch of length $\aleph_1$?

Aronszajn, in the 1930s, found a counter-example, which he communicated to Kurepa.

Suslin, Aronszajn and Kurepa then compared the cardinality and structure of their trees with cardinality and the structure of the main known sets.

Depending on the set theory and the axioms that characterize them, the existence of certain trees is possible or not possible.

2 Suslin's problem is the following: given a non-empty totally ordered set R with the following four properties: (1) R does not have a least nor a greatest element; (2) the order on R is dense (between any two elements there is another); (3) the order on R is complete, in the sense that every non-empty bounded subset has a supremum and an infimum; (4) every collection of mutually disjoint non-empty open intervals in R is countable (this is the countable chain condition for the order topology of R): is R necessarily order-isomorphic to the real line $\mathbb{R}$?
The Suslin hypothesis says that there are no Suslin lines: that every countable-chain-condition dense complete linear order without endpoints is isomorphic to the real line. An equivalent statement is that every tree of height ω_1 either has a branch of length ω_1 or an antichain of cardinality $\aleph_1$.
The generalized Suslin hypothesis says that for every infinite regular cardinal κ every tree of height κ either has a branch of length κ or an antichain of cardinality κ. The existence of Suslin lines is equivalent to the existence of Suslin trees and to Suslin algebras.
The Suslin hypothesis is independent of ZFC. Jech (1967) and Tennenbaum (1968) independently used forcing methods to construct models of ZFC in which Suslin lines exist. Jensen later proved that Suslin lines exist in Gödel's theory $V = L$. On the other hand, Solovay and Tennenbaum (1971) used forcing to construct a model of ZFC in which there are no Suslin lines; more precisely, they showed that Martin's axiom plus the negation of the continuum hypothesis implies the Suslin hypothesis.
The Suslin hypothesis is also independent of both the generalized continuum hypothesis (proved by Ronald Jensen) and of the negation of the continuum hypothesis. At the present, it is not known whether the generalized Suslin hypothesis is consistent with the generalized continuum hypothesis.

Examples

1) Suslin trees (height $\aleph_1$, countable branches, countable antichains);

2) Aronszajn trees (height $\aleph_1$, countable levels);

3) Kurepa trees (height $\aleph_1$, countable levels, at least $\aleph_2$ uncountable branches);

4) etc.

Among the infinite trees, infinite hierarchies and classifications naturally appear. But concerning the possibility that they actually exist in the real world, mathematics does not bring more certainty here than for partial orders.

3.5.2. *Ordinal infinite classifications*

THEOREM 3.10.– *Let A be an infinite set of ordinals, A' an infinite part of A and $[A]^n$ a subset of n elements of A. An ordinal infinite classification exists if there exists:*

$$\phi : [A']^{n-1} \to [A]^n$$

such that, for two elements s and t of $[A]^{n-1}$ with $s < t$, there exists $t' = \phi(t)$, such that $\phi(s) < t'$.

There exist, however, some obstructions. For example, Erdös (1956) proved that:

$$[\omega]^\omega \nrightarrow [\omega]^{\omega+1}$$

REMARK 3.3.– *The existence of ordinal infinite classifications (or chains of well-ordered partitions) is not always possible. There too, the passage to infinity does not allow any assurance of the existence of these structures (on all that, see Parrochia and Neuville 2013, pp. 241–260).*

3.6. Moral problems in a finite world

Now consider the finite set of human actions. Human behavior is controlled by what we call "Morals" or "Ethics", which generally include values and an ordering of values.

Values define things some men consider to be good, desirable or important. There exist various types of values such as personal values (which guide our perception of right and wrong), cultural and social values (such as gender and race equality), then professional values (which define our work ethics). Values such as loyalty, courage and honesty often carry currency in a professional environment. Generally, we call "moral values" the behavioral practices, goals and habits which are validated by the society we are part of. This set of values typically becomes embedded in our behavior through a long process of observation, education, conditioning and social guidelines.

A first problem is that of the universality of these values and, assuming that we agree on them, that of the universality of the classification (usually a hierarchy) of these values.

A second problem is that of the relationship between moral values and what could be called "core" values: moral values are concerned with right and wrong, and also define what is socially acceptable as good or evil. "Core values" are rather a set of things and beliefs that a person considers central to their identity. As well, there is often overlap between a person's core values and moral values. For example, loyalty and commitment could be both core and moral values. To quote an example, the commitment to environmental sustainability is a core value that increasingly dictates the personal and professional beliefs of people and organizations. Core values are often noted and rewarded in the professional sphere. For example, someone who is naturally hardworking values discipline is likely to do well at work and advance in their career.

Mathematics can be of some help in solving these kinds of problems. For example, it is always possible to compare "subjective classifications" hoping to arrive at an "objective order" (see Bartozsynski 1971). But, as we have shown elsewhere (see Parrochia 2004, pp. 193–201), the possibility of constructing this rests on axioms which are not insignificant.

The question of the superposition of two different hierarchies (e.g. that of moral values and fundamental values, which can very well become entangled) can also be settled by appealing to mathematics. Given two chains (finite or countable), we have seen previously that we can define what is called a *bichain*, each chain being a part of the partition of it. But we have also seen that every finite or countable bichain is included in a particular infinite countable bichain called "universal bichain" (see Pouzet 1978).

Let us now see things a little more concretely, on the example of a particular moral philosophy.

For some philosophers – Max Scheler, for example – the ordering of values is a strong hierarchy, i.e. from a mathematical point of view, a *chain*: it begins at the lowest level with "virtual values" (A). Then come "vital values" (B), "spiritual values" (C) and the top of the hierarchy is formed by the values of the Holy (D). This is an instantiation of a total order (see Figure 3.2 (left)).

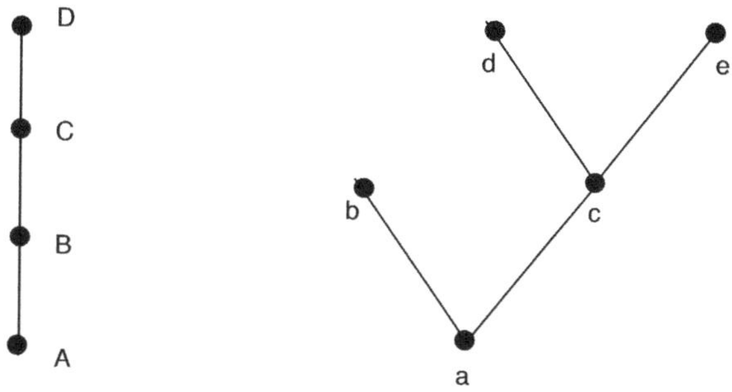

Figure 3.2. *Hierarchy and partially ordered set*

But many problems arise:

Firstly, such a hierarchy is it a universal one? If not, should we weaken this hierarchy and substitute for it a partial order – for example, a simple classification or a preorder? But since Aristotle, we can see that there is no consensus among philosophers on the existence of a non-questionable typology of values.

Secondly, although we admit the hierarchy of Scheler (or any other similar hierarchy), the problem of *moral dilemmas* remains.

Suppose, indeed, the following situation: my brother and his girlfriend are drowning. I can swim, but I cannot save both. Should I save my brother or girlfriend? Should I save my brother because he is my brother, the person closest to me, or his girlfriend because she is a woman, i.e. someone to whom, in principle, we first have to assist.

Similar situations can be reproduced at any moment of life: in *existentialism is a humanism*, Sartre describes the case of a student who hesitates between joining the French forces in exile to fight the German enemy, avenging his father there, and staying with his mother to help her live.

But we can easily find many other examples. Let us imagine this one: Mr X is married but he had an affair with another woman. If Mr X conceals from his spouse that he has had an affair, he will act dishonestly; but if he confesses to her, she risks leaving him and the divorce breaks the family.

Now consider this last dilemma: Ms. Y's doctor is diagnosing terminal cancer in his patient, whom he has known for a long time. Given the depressive state of the latter, he considers that she would not bear to know she is suffering from this disease. His duty to inform his patient of her condition conflicts with that of not destroying what is left of hope in life. Should he tell her the truth or hide her condition?

In the example of Mr X, the values of honesty and family solidarity conflict because they suggest to act in two different ways: if the husband acts according to the value of honesty, he will confess his adultery; but if he acts according to the value of family solidarity, he will not admit it.

In the last situation, a moral principle and duty suggests adopting opposing behaviors: the doctor's duty not to lie is contrary to the principle of beneficence towards their patient.

Without being chaos, the moral life can confront us with inextricable situations because, the most of the time, we do not know what is the well-ordering of our actions. Sometimes, they cannot be ordered at all, but only preordered or partially ordered: for example, I can know that an action a must take priority over b and c shares, and c priority over d and e. But I may be unable to compare b and c, on the one hand, and d and e, on the other hand, if the value system I obey is not a hierarchy (see Figure 3.1 (right)). This situations generate what are called "dilemmas".

It is in the Middle Ages that we find the first discussions of the question of whether moral dilemmas are possible. The philosophers' response is generally negative. Neither Thomas Aquinas, nor Kant, nor Mill admit the possibility of moral dilemmas, because their existence would mean that the moral theory in which they appear is inconsistent. It seems that it is especially among contemporary philosophers that we find voices that, following Lemmon and Williams, defend moral dilemmas (see Lemmon 1962; Williams 1966; Williams and Atkinson 1966).

The major argument in favor of the non-existence of dilemmas is that they make the underlying moral theory inconsistent. After the developments of the deontic logic, we can show that very clearly. Let us recall here that two basic principles regulate the classical systems of deontic logic:

1) the Kantian principle that "must" implies "may" (if an agent has the obligation to do something, they can do it);

2) the agglomeration principle (if an agent has an obligation to do one thing and also has an obligation to do something else, then they have an obligation to do both).

Now, say a and b are action names. O(a) means that the action a is obligatory and M(a) means that it is possible to do a. If A and B can be substituted by expressions referring to actions, such as a and b, we can formulate the principle that "must" implies "may" as follows:

$$(MM) \qquad O(A) \supset M(A).$$

The principle of agglomeration, meanwhile, can be symbolized as follows:

$$(AG) \qquad O(A) \wedge O(B)) \supset O(A \wedge B).$$

Now, we can show that, from these principles and propositions describing moral dilemmas, we obtain a paradox. Indeed, if there are moral dilemmas, the following three propositions are true:

(1) $O(a)$;

(2) $O(b)$;

(3) $\neg M(a \wedge b)$.

Suppose the agglomeration principle (AG) is correct. We then obtain:

(4) $(O(a) \wedge O(b)) \supset O(a \wedge b)$.

From this principle and from (1) and (2), we arrive by simple propositional calculation to:

(5) $O(a \wedge b)$.

Let us admit that "must" implies "may" (MM principle), and apply it to (5). We obtain:

(6) $O(a \wedge b) \supset M(a \wedge b)$.

From (5) and (6), we can then obtain:

(7) $M(a \wedge b)$.

Now, if we put (7) and (3) together, we get the following contradiction:

(8) $M(a \wedge b) \wedge \neg M(a \wedge b)$.

(8) is a marked contradiction. Therefore, we must either give up moral dilemmas or reject (at least) one of the principles that allow us to infer the contradiction. Some philosophers – Lemmon, for example – opt for abandoning the principle that "must" implies "may", while some others – Williams and van Fraassen, for example – prefer to reject the agglomeration principle (Lemmon 1962 p. 114, note 2; pp. 118–119; Fraassen 1973, p. 13). What is the plausibility of these principles?

The intuition that is expressed by the principle that "must" implies "may" is that we cannot require someone to do the impossible: it just seems too hard to require someone to do anything that they cannot, with the best will, accomplish. However, there seem to be counterexamples to this principle. Thus, some have argued that a creditor who is no longer able to pay these debts still has an obligation to do so (see Stocker (1971, pp. 95–96), and Brink (1989, pp. 230–231), for an answer).

As for the agglomeration principle, it also has a certain intuitive plausibility: in ordinary cases, if an agent has two obligations, it is clear that the agent must perform both the first action and the second. Indeed, it does not seem sufficient to say that the agent has the choice between the first and the second action. Despite this, it is not clear that, in the case of insoluble conflicts, it is true that the agent has a conjunctive obligation. Take the example of two people drowning and whose agent can only save one of them. It can be argued here that the only obligation involved is disjunctive: what is morally required is only for the agent to save one or the other of the two persons.

The existence of moral dilemmas leads to the rejection of one or the other of the MM and AG principles, or both. And since, by hypothesis, we cannot prioritize values or obligations according to a total order, the existence of these dilemmas in the daily life of men seems absolutely unavoidable. The result is that all the actions of men are best ordered in a partial order or a preorder, but certainly not in a total order.

3.7. Order versus circularity

The previous reflection shows how much the world of thought, philosophical and scientific, has been marked by order and hierarchies. A question arises, however: is an unordered (non-hierarchical) universe (see Finsler 1926; Aczel 1988) possible? Unexpectedly, the answer is "yes".

The theory of Aczel's hypersets (which takes up insights from Finsler) has proved its (relative) consistency. Finsler's or Aczel's theories are no longer consistent (or inconsistent) than Zermelo–Fraenkel's set theory. We have:

Cons ZFC $\Rightarrow$ Cons ZFC + AAF

where AAF being the axiom of anti-foundation.

This result makes possible a totally non-hierarchical vision of world and a new conception of the universe of sets. Let us first recall what is the axiom of foundation.

In set theory, the foundation axiom (AF), introduced by Von Neumann, was intended to prohibit "circular" sets (like the Mirimanoff set), i.e. paradoxical sets that can manifest as sequences:

$$S = \{a_0, a_1, ..., a_n, ..., a_0\} \; |n < \omega \text{ such that } a_0 > a_1 > a_2 > ... > a_0.$$

The foundation axiom forbids "circular" sets of this type. Thanks to this axiom, the only sets accepted in set theory are "well-founded" sets, i.e. the usual non-circular sets.

Indeed, circular sets exist. In Finsler (1926), we can find four examples of this kind of set:

1) $J = \{J\}$ (named the J-set);

2) $A = \{A, B\}; B = A$ (named "Fibonacci–Finsler sets");

3) $N_0 = 0; N_{n+1} = \{N_n\}$;

4) $M_n = \{N_n, M_{n+1}\}$.

To these ones must be added Aczel's graphs: the hyperset theory systematizes the use of rooted graphs and trees introduced by Finsler (1926, 1963).

For example, the Fibonacci–Finsler set can be represented in two ways, depending on whether it is rooted in A or B. We have (see Figure 3.3):

$$B = \{A\}A = \{A, B\};$$

$$A = \{A, B\}B = \{A\}.$$

Can we get rid of the notion of order? To this question, a rational person will tend to answer negatively. Everything suggests that the world around us is hierarchically ordered and that it exists, as we say, "an order of things". Thus, I am a professor in a faculty, which is itself located at a university, and is part of a group or a regional pole (Rhône-Alpes), which itself constitutes a subset of all French universities.

Descartes once counseled that it was better to change our desires than the order of the world, and after several bloody revolutions since the 18th century, which, on reflection, do not seem to have substantially modified this order, we would be tempted to say he was right.

Figure 3.3. *Fibonacci–Finsler sets*

However, as curious as it may seem, mathematics allows for another world view, where people and objects are taken in relations that are different from membership or inclusion, or, more accurately, in relations that have taken a "circular" form.

The facetious Aczel gives a funny example of that by plunging the names of authors of set theories with a very hierarchical flavor (Zermelo, Fraenkel, Hilbert) in a circular set that questions the relevance of their subject. This may be formally defined this way: let a, b, c be the following sets:

$$a = \{b, c\}; \qquad b = \{Zermelo, Fraenkel\};$$

$$c = \{Hilbert, Fraenkel, b\}.$$

This burlesque situation is described by the graph in Figure 3.4.

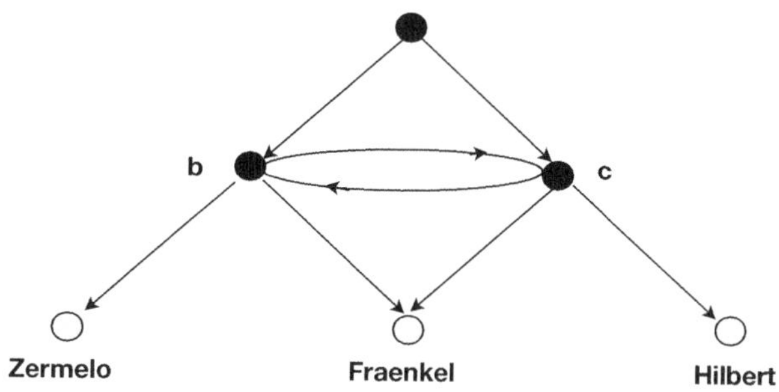

Figure 3.4. *Circular theoricians*

The logic of nature, like that of economics or technology, seems to oppose such reveries. Large fish eat small, large companies have a whole hierarchy of subcontractors and manufactured objects themselves seem to be built or assembled in a certain order, total or partial. What would be a world where these laws would no longer be respected? The thing seems impossible: if a pipe A fits perfectly in a pipe B which fits itself in a pipe C, we do not see that this pipe C can fit into A (see Figure 3.5). Could it moreover, that it would not serve much: in principle, a pipe is made to convey material flows from a starting point to a point of arrival, not to turn them around like a snake that bites its tail.

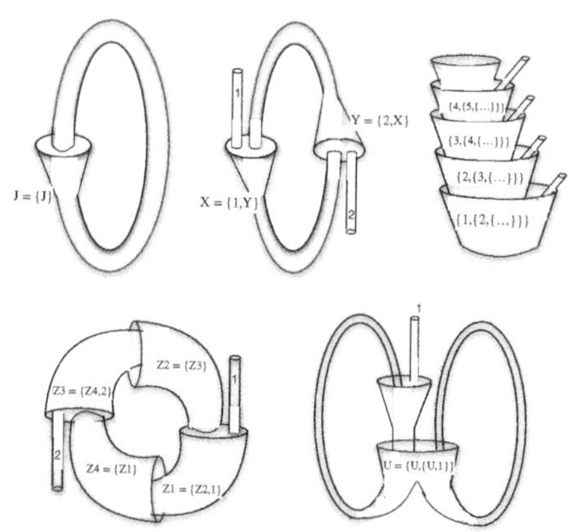

Figure 3.5. *Circular objects (after Delahaye 1994). For a color version of this figure, see http://www.iste.co.uk/parrochia/graphs.zip*

However, without considering a geometry for crazy plumber, we can imagine situations of this kind: nothing prevents, for example, a highway, to lead to a smaller road, then to an even narrower path, which finally returns on the departure route. And sometimes, by complicating an already difficult problem, we finally find how to solve it. In nature as in society or in the world of technical objects, circularity exists: it suffices to widen its field of vision to escape the sad prospect of the military organization to which laziness impels us to adhere.

Art, fortunately, sometimes allows us to escape from this vision of things. Among the works that open the imagination to another world, we can mention, in particular, the works of the engraver Escher, which often very precisely translate this vertiginous circularity in which we are caught (see Figure 3.6).

Figure 3.6. *M.C. Escher's Print gallery: an illustration of the set* $J = \{J\}$ *(source: © 2022 The M.C. Escher Company-The Netherlands. All rights reserved. www.mcescher.com)*

3.8. Conclusion

The order, which can be partial or total, as well as strict, wide, linear, well-order or simple quasiorder, or even better order or better quasiorder, is said in different directions in mathematics.

The relations between ordered structures, pre- or quasiordered, partially or totally, are complex, and do not always result in the existence of possible or real structures.

Consequently, philosophers (as well as scientists) must limit their ambitions. The world is not only more confused than it seems. It is actually also very restive to order.

We can therefore easily prefer circularity (Finsler, Aczel) or leave it, after all, to its disorder if something better is not possible.

4

Towards a Formal Philosophy

The idea of describing and understanding philosophical structures through formal models goes back to Boole and his logical algebra, which allowed him to re-write parts of the Spinoza and Clarke systems. But this idea grew in importance through the various attempts to formalize the Hegelian dialectics (see Trendelenburg (1840) and Dubarle and Doz (1972)). The oldest attempts were analyzed by Apostel (1967). Various logics were used, in particular temporal logics (see Rogowski 1964; Gardies 1975) and, more generally, the reflection on dialectic as a kind of logic. This has been the theme of many symposia or articles. Among others, see Guillaumaud (1980) and, for a different opinion, see Beth (1948) or Popper (1940). Various essays, such as those of Guenther (1964) or Kosok (1966, 1972), have already been identified in an earlier article by Gauthier (1967). In a study of Hegel, we ourselves have discussed these propositions (see Parrochia 1990), as well as more recent attempts (see Marconi (1979) or Arruda et al. (1980)), such as those of paraconsistent logic, and have also proposed new formalisms, likely to be applied to different philosophies.

We would now like to return to the essay by Dubarle (see Dubarle and Doz 1972), which we think, on reflection, inaugurates, beyond the formalization of the Hegelian system, a new way of conceiving the relationship between philosophy and formal structures, to which we would like to give, in the context of an old work (see Parrochia 1993a), a certain extension.

4.1. Asenjo's systems and Dubarle's formalization of Hegelianism

We will start here with a short article by the mathematician Asenjo (see Asenjo and Tamburino 1975), a specialist in antinomic calculations and paraconsistent logic (see Asenjo 1966, 1965), to whom he returns to lay the foundations for a precise formalization of the dialectic of which the project by Dubarle is, in actually, a variant. Distinguishing two types of dialectical reasoning (the monstrous, who deliberately violate the principle of contradiction, and the others, who respect it), Asenjo shows that the Hegel does not intervene as a rule of inference but as a rule of formation of statements. He builds on this basis a calculation of the predicates of a higher order than the dialectic, and of which this becomes a calculus in which the famous *Aufhebung* is carried by two rules called "enveloping rules".

4.1.1. *Asenjo's systems and Dubarle's case*

This calculus could in principle be infinite, but with the Hegelian system being finite (which Asenjo interprets from a formal point of view by saying that there are logical categories that have no negation and, from there, can no longer be synthesized (Asenjo and Tamburino 1975), we can confine ourselves to defining a finite number of predicate letters. Any calculus of predicates leading to a kind of number theory, the author then develops a system that is presented as an extension of the proposed formalization. This formalization is introduced in a normal way as a calculus, comprising an alphabet:

– individual variables $x_1, x_2, ..., ...$;

– individual constants $R_1, R_2, ...$;

– function letters;

– predicate constants $A_{11}, A_{12}, ..., S_2, S_3, ..., S_k, ..., A_{2k+1}, A_{31}, A_{32}, ...$;

– predicate variables;

– propositional connectors $\neg, \supset, ...$;

– quantifiers;

– parentheses and commas;

– eleven training rules, two of which are really specific and want to translate the basic aspect of the calculus:

R6: t_1 and t_2 are terms, and R_i is an individual constant, so $S_2(t_1 R_i, \neg t_2 R_i)$ is a well-formed formula (abbreviated as $S_2(t_1 t_2 R_i)$ (see Asenjo and Tamburino 1975).

R7: if P_j is a well-formed formula enveloping only predicate letters $S_2, ..., S_j$ $(j < k)$, then $S_j + 1(P_j, \neg P'_j)$ is a well-formed formula, where the "prime" of P'_j indicates that the terms and, in particular, the individual variables in P'_j, are different from those of P_j, although P_j and P'_j have the same individual constant (see Asenjo and Tamburino 1975).

R6 thus defines the dialectical opposition and R7, as we have just seen, the "Aufhebung" or "supersumption".

The arithmetic interpretation of this formalism is then as follows. Given a set of numbers S (domain of interpretation of the individual variables) and the constant R_1 associated with a subset of S, the well-formed formula (wff) $S_2(x_1, x_2, R_1)$ can be satisfied by certain pairs of numbers and not by others. If, for example, $S = \{0, 1, i\}$ and $R_1 = \{0, 1\}$, then the dialectical numbers are (0, i) and (1, i). S_2 is a binary predicate. For a predicate S_j with j places, we will have dialectical numbers with $N(j)$ components, with the odd place components belonging to R_1, and the even places being filled by numbers from the complementary of R_1 in S. Thus, if R_1 is the set of real numbers, the complement of R_1 in S is the set of pure imaginary (noted R_i). The dialectical numbers are then of the form:

$$(a_1, a_2 i), (a_1, a_2 i, a_3, a_4 i), ..., (a_1, a_2 i, ..., a_{N(k)} i).$$

On the binary dialectical numbers, we can define the operations:

$$(a_1, a_2 i) + (b_1, b_2 i) = ((a_1 + b_1), (a_2 + b_2) i)$$

$$(a_1, a_2 i).(b_1, b_2 i) = ((a_1.b_1), (a_2.b_2) i)$$

where the addition is similar to that of the complex numbers, but not the multiplication. More than an actual formalization, it is, as we can see, a system capable of generating a whole class of formalisms, according to the meaning we will give to the set S and R_1.

We would now like to demonstrate that Dubarle's system can easily be interpreted as a variant of Asenjo's class of formalisms.

Rule R6 of the Asenjo system is replaced by rule R'6 by changing the formula:

$$S_2(t_1 R_i, \neg t_2 R_i)$$

into the formula:

$$S_2(t_1 R_i, t_2 R_i).$$

We then ask:

$$R_1 = S = \{0, 1\}$$

Hence, the set of numbers obtained is $B^2 = \{(0,0), (0,1), (1,0), (1,1)\}$, which, provided with the laws of addition and multiplication defined by Asenjo, is just the Boolean ring.

4.1.2. *Dubarle, Parmenides' thought and Hegel*

Dubarle, in fact, proceeded differently, starting from the analysis of Parmenidian discourse. To think of the very existence of any reality whatsoever, it is necessary logical to posit the existence of this reality. At the origin of philosophy, this is what Parmenides does by striving to think about the totality of reality under a concept: the concept of being. But, he immediately realizes that in order to hold a discourse on this concept, he is obliged to differentiate it from another, its negation, which is the term for nothingness or non-being. In a logic of the binary type, we will say that, for there to be a 1, an indexed value that is retained as the truth value of the expressions defined in this logic, there must be a zero. If there is "true", there must be "false". It is of the very essence of thought that we can only think of something by opposing it to something else, and therefore cannot think of the whole, or the universal class, contrasting it with "nothing", or the empty class. It is, according to Dubarle, the very discovery of Parmenides.

Hegel, on the contrary, tried to interiorize the Absolute itself and this opposition – which stood in Parmenides between the Absolute and the Negation of the Absolute, in other words, Nothingness – is relativized. This nothingness then becomes a simple fact of negation interior to the concept, whose philosophical meaning is the idea of a certain energy of freedom specific to this concept, which is exercised precisely as negativity, as opposition to what is. As a result, the term null, which should have symbolized the suppression of the universal, is purely and simply erased. Hegel cannot represent it as a Parmenidian nothingness, an abstract zero, without putting it outside the Absolute, and at the same time, he has no formalism capable of translating it otherwise. According to Dubarle, using the Boolean ring $B^2 = \{0, 1\}^2$ makes it possible to solve the problem.

Following Dubarle, we then associate with each pair of numbers in B^2, respectively, the constants of structure Λ, which corresponds to the null term of the Boolean algebra, with U standing for Universal, P for Particular and S for Singular (the three moments of the concept plus the empty term (Λ), necessary to define the negation). The addition operation of the commutative ring is then normally interpreted by the symmetric difference $\backslash$, and the operation of multiplication by the intersection.

Two cases are interesting: the *Negation* in the Hegelian sense, or passage from the universal (U) to the pair formed by the particular and the null term (P, Λ); the *Aufhebung*, or passage of this pair (P, Λ) to the real singular (S). Two operations are necessary each time:

1) for the negation: the transition from U to Λ is done by taking the intersection of U and P (deposition operation). The transition from U to P (complementation) corresponds to the intersection of U and S in the ring;

2) for the supersumption (Aufhebung): the transition from Λ to S is again complementation, but this time corresponds to the symmetric difference between Λ and S in the ring. The transition from P to S ($P \setminus S = S$) is called "raising". The tables of operations of the ring can be easily established (where $\setminus$ traditionally represents the symmetric difference and $\cap$ is the set intersection):

$\setminus$	Λ	P	U	S
Λ	Λ	P	U	S
P	P	Λ	S	U
U	U	S	Λ	P
S	S	U	P	Λ

$\cap$	Λ	P	U	S
Λ	Λ	Λ	Λ	Λ
P	Λ	P	Λ	P
U	Λ	Λ	U	U
S	Λ	P	U	S

This four-element product ring defines the base of the system. But this is, as we know, the structure of a hierarchical tree. In the model of Dubarle, the general structure is transmitted in relation to each level (see Dubarle and Doz 1972, p. 128). For a universal U, for example, we can take again the universal of this universal (noted U, U), the particular of this universal (U, P), etc. The Hegelian system thus comprises virtually 4^n terms (n = number of levels).

4.1.3. *Projective algebras*

This beautiful structure makes it possible to consider the Hegelian system, not exactly as a "circle of circles", which is the Hegelian metaphoric formula, but truly as a *ring of rings*, in the mathematical sense of the term. Since Boolean algebras have good correspondences in lattice theory and topology, we could also show that the Hegelian system – if it is, at each of its levels, a Boolean ring – is also a lattice, and a discrete topological space: it is easily translated by duality, using the Stone isomorphism (see Ponasse and Carréga 1979).

At this stage, however, the Hegelian logic has only been represented by the trace left by conceptual operations in the finite understanding. But from the viewpoint of the Hegelian Reason, the concept must express the infinitude of the "Thing in itself". The only way to reconcile the finiteness of the linguistic expression and the infinitude of what is expressed is then to consider the formalism we have constructed as the trace, in the finite, of relations existing in the infinite.

4.1.3.1. *The case of Parmenides*

Note that this question is already encountered in Parmenides. As Parmenidian thought is of the binary type, it is defined on a simple Boolean set $B = \{0, 1\}$. Firstly, we can consider any Boolean ring $B = \{0, 1\}$ as a vector space over the field U with two elements. This Boolean algebra must thus be constituted in *projective algebra*, according to the methods used in finite projective geometry (on this point, see Monjardet (1968), or, for a more technical viewpoint, see Dembowski (1968) and Hirschfeld (1979)). We can therefore form a projective space deduced from the vector space, and take a system of homogeneous coordinates as the basis of this space, in other words a system expressed with the very elements of the field. We then obtain the one-dimensional geometry PG[1,2], in other words, the projective line that has the following form.

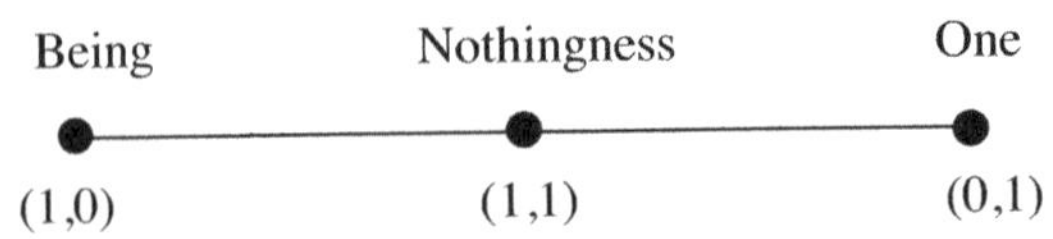

Figure 4.1. *The projective geometry PG[1,2] or Parmenides' way of thinking*

Here, we have two "finite distance" points (1,0) and (1,1) and an infinite distance point (0,1), of which these two are the projections. In a projective writing, we would rather note these points 0/1,1/1 and 1/0, but it does not matter if we choose another writing. The collinearity of the points (their alignment) results in a simple algebraic relation: if we introduce an operation $\dot{+}$ called the operation of "transfer" on the set of pairs, defining the transfer by taking the symmetric difference term at term between the elements of the pairs, then, we have, between the finite and the infinite, a relation of the following kind. For all a, b, c, d belonging to the Boolean ring, we have:

$$(a, b)\dot{+}(c, d) = (a + c, b + d)$$

This symmetric difference is not exactly the same operation as that which is defined between the elements of the finite set. It is, in this context, a transcendent operation. According to Dubarle, this projective line corresponds exactly to the modeling of Parmenidian discourse. Parmenides opposes two finite entities (nothingness and being), as well as a totalizing one, which is Being with a capital letter, or the One, which encloses an infinite comprehension in itself.

4.1.3.2. *The case of Hegel*

In the Hegelian case, the starting point is no longer the field $B = \{0, 1\}$ but, as we have seen, the product ring $\{0, 1\}^2$, isomorphic to the algebra with four constants of structure. As we know, a Boolean algebra is normally a vector space on the field K with two elements $\{0, 1\}$. On the product set $\{0, 1\}^2$, which is a ring, it is therefore a module. Anyway, we can still build a projective geometry on it. For example, for the four constants, we will take the following coordinates, which are therefore those of the finite distance points: (1,0,0) for Λ; (1,1,1) for S; (1,0,1) for U and (1,1,0) for P.

The collinearity of these points at a finite distance and of their respective projections into infinity is calculated in the same way as previously. We will also use an operation $\dot{+}$, which, besides its immanent meaning in the ring, will here take on the following transcendent one: we will have, for all $a, b, c, d, e, f \in B^2$:

$$(a, b, c)\dot{+}(d, e, f) = (a + d, b + e, c + f).$$

We finally obtain a representation of the four basic terms in a well-known configuration: the seven-point projective plane or Fano plane (see Dubarle and Doz 1972, p. 179). The elements Λ, P, U, S become the finite projections of their respective infinite representations (the Universal, the Particular, the Singular). The construction could be apparently generalized to the product sets, as Dubarle shows (see Dubarle and Doz 1972, pp. 185–186)[1]. We find here, in Figure 4.1, covered with Hegelian concepts, the Fano configuration already represented in Figure 2.28 (left).

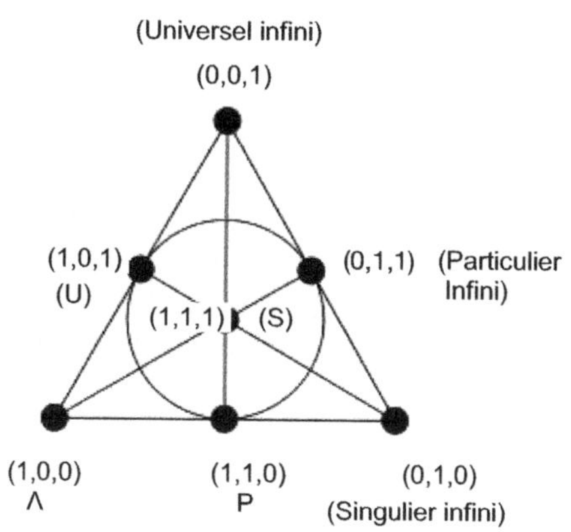

Figure 4.2. *The plane PG[2, 2]*

1 Dubarle speaks of an unlimited sequence of projective geometries of the type PG[2^n, 2]. In the current notation, for example, that of Hirschfeld (1979), it would be better to write PG[2, 2^n], with, for $n = 1$, the configuration of Fano or PG[2,2]. Remember that the number of the levels of the system does not exceed $\ell = 6$.

4.2. Some criticisms

This beautiful clarification of the Hegelian approach, without doubt, has several criticisms:

1) the structure is too regular to really apply to the detail of the Hegelian text: if only in the logic, we would identify the existence of asymmetries (two or four moments instead of the usual three). How do we report exceptions?;

2) the transformation of triplicity into quadruplicity, which seems to be based on a Hegelian text, is not absolutely convincing. It is obviously called by the necessities of formalism and generates phantom positions, at the rate of $4^\ell - 3^\ell$, with ℓ being the number of levels of the Hegelian tree. As, in some developments, we can have $\ell = 5$ or $\ell = 6$, the number of phantom positions can reach a number between 781 and 3,367;

3) the interpretation of negation as Aufhebung in terms of Boolean operations (complementation and symmetrical difference) undoubtedly gives rigidity to the text. The existence of a multiplicity of different dialectics in Hegel seems, in fact, to rest on a much more flexible, far more local conception of negation, and therefore of complementation. This one, moreover, relates less to terms than to sets of semantic determinations;

4) the formalism by Dubarle gives no explicit method that makes it possible to extract the determinations U and P from any concept, of which, moreover, we have never, hic and nunc, the state S;

5) the successive instantiations of projective geometries on each level of the system pose a problem. At a level ℓ, let ν be the number of points at a finite distance, and μ be the number of points at infinity. Knowing that the total number of points of a projective plane PG$[n, q]$ is equal to $m = n^2 + n + 1$, and that $\nu = n^2$ (the number of moments or conceptual positions), we can deduce the following results (see Table 4.1).

ℓ	$\nu = n^2$	m	$\mu = m - \nu$	PG$[n, q]$
1	4	7	3	PG[2,2]
2	16	21	5	PG[2,4]
3	64	73	9	PG[2,8]
4	256	273	17	PG[2,16]
5	1,024	1,057	33	PG[2,32]
6	4,096	4,161	65	PG[2,64]

Table 4.1. *Characteristics of "Hegelian" projective planes*

As we can see, beyond the level $\ell = 1$, it is not certain that we can really attribute a Hegelian meaning to the numerous points at infinity which multiply in later projective geometries.

Despite its limitations, such formalization remains, at a certain level, satisfactory: whether it is real Hegelianism or, more likely, as we think, a regularized variant of it, it highlights a possibility of philosophical thought that the formal models make it possible to compare with others, even to locate in a classification. It is this new perspective that we will retain here.

4.3. Porphyry and the neoplatonist mode of thought

Dubarle conceived Hegel's logic as an extension of Parmenidian thought, which he saw as inaugurating a binary mode of thought whose Hegelianism was only a complexification.

But if Parmenides is, indeed, at the origin of the thought to which modern science and technology owe so much, we will notice that another form of thought, resulting from this one, but in rupture with it, was already profiled since Neoplatonism. In Porphyry, in particular, more than in his master Plotinus, the fundamental opposition of reality does not pass between a single being and a non-being, since, as Hadot writes, it is only in reality that there are modes of being as non-being (see Hadot 1968, p. 177). The main opposition is then between this cleaved reality and the non-being above the being. But, as Hadot remarks, "the non-being above the being is also only a 'projection', as the commentary on the *Parmenides* shows[2]. If God appears to us as no-because, it is because we are different from Him, so nothing in relation to Him. It is He who is the only true Being and in relation to Him, we are non-beings" (see Hadot 1968, p. 1, 177). The difference with binary thinking is that there are several types of entities here, such as, for example, the non-being conceived as deprivation of true reality (NB1), the non-being according to the power (NB2), the non-being according to the alterity (NB3) and finally the non-being above the being (NB4).

We must therefore think, next to the projective Parmenidian geometry, as described by Dubarle, what we may call a *projective geometry of reality*, perhaps closer, in some respects, to Hegel's philosophy, and probably also – we are only

2 Porphyry, *On the Parmenides,* IV, 26: "It is therefore we who are the Nothingness with respect to Him, as for Him, He is the only true being (if we mean it in the sense that I say it) in relation to all things that are after Him, not bearing any relation to these things and not turning away from the solitude that is proper to Him to experience relationship and multiplicity ..." (see Hadot 1968, p. 2, 77).

suggesting it here – of all thought that rejects a simple opposition between being and nothingness.

We believe that such a way of thinking is capable of illuminating both theological and philosophical thoughts on some of their most profound operations.

What is at stake here is, at once, the problem of the conception of nothingness or, more generally, of negation, in the thought of who must at the same time endeavor to explain its origin, and the origin of this origin, in its relation with the universe. No one better than Emile Bréhier has been able to define more pertinently the rather paradoxical aspect of these questions in Greek Neoplatonism:

> the origin cannot, as such, possess any of the characters that beings possess to explain and to deduce; for it would then be one thing among the other things, a being among other beings. But, possessing no character of beings, it appears, to thought that would like to grasp it as a pure non-being, a nothingness of being. On the other hand, insofar as it is the source of reality, it must possess a mode of existence superior to that of any being accessible to thought; it is therefore a kind of nothingness that is superior to existence (see Bréhier 1919; Koyré 1929, p. 303).

To the existence of beings are thus opposed two kinds of non-being: the one that is the absolute negation – pure and simple – and the one that is the origin and the source of being, which poses, of course, the most difficulties:

> as soon as you try to determine it and reach it by thought, you make it a being, and from that time it is no longer origin; and because it is a being, one must ask again what its origin is. If, on the contrary, you leave it fully indeterminate, it appears no different from pure non-being, and consequently it is no longer the origin of being. You must at the same time ask and withdraw determinations accessible to the thought, ask them so that the origin is not a pure nothingness and withdraw them so that it is really origin and source and not only a term of this reality (see Bréhier 1919).

The Hegelian resolution of this paradox consists of considering each time negation as a local determination – in other words, as a provisional pseudo-complement. The contradiction is thus reduced to the state of a simple determination and the system finishes by resolving the antinomy by instituting a progressive determination of the most indeterminate nothingness (the abstract being of the beginning) which leads to this negation of all finite reality that is the Absolute.

4.4. A variant of Dubarle's formalism

We must now transpose the ideas of Dubarle onto this new mode of thinking. Dubarle built a Boolean algebra that admitted a projective expression. To get a mathematical model of the Porphyrian thought, while keeping the same style of formalization, we must now, as we can guess, construct a basic algebra that translates the constraints mentioned above. In Porphyry, as we have seen, the modes of non-being are three in number, to which must be added the only true being who really exists, and who is God, all the rest being only projections. In the finite, the positions are pseudo-complements and we pass from one position to another by a process which can be perfectly described thanks to a certain kind of algebra.

Using a possibility described in our previous article (see Parrochia 1990), we introduce the following formalism: we call "Heyting's algebra" or "pseudo-complemented algebra" an implicative lattice, in other words, a partially ordered algebra, with elements 0 and 1, and three binary operations $\wedge$, $\vee$ and $\rightarrow$, such that we have the following axioms:

1) $a \wedge b \leq a$;

2) $a \wedge b \leq b$;

3) $c \leq a$ and $c \leq b \rightarrow c \leq a \wedge b$;

4) $a \leq a \vee b$;

5) $b \leq a \vee b$;

6) $a \leq c$ and $b \leq c \rightarrow a \leq a \vee b \leq c$;

7) $a \wedge (a \rightarrow b) \leq b$;

8) $a \wedge c \leq b \rightarrow c \leq a \rightarrow b$.

If, now, axioms 1–6 being maintained, we have, instead of 7–8, the following axioms:

9) $a \leq b \vee (a - b)$;

10) $a \leq b \vee c \rightarrow a - c \leq b$;

we have, instead of an implicative lattice, a subtractive lattice (or Brouwer's algebra).

To the pseudo-complement $a \to b$ of the implicative lattice (least upper bound of the class of x satisfying $a \wedge x \leq b$) here corresponds the Brouwerian complement, in other words the greatest lower bound of the x satisfying the relation $a \leq b \vee x$. In a Heyting algebra, the complement of a is written as $a \to 0$ or a^*. In a Brouwer algebra, the Brouwerian complement is denoted as $1 - a$ or $\neg a$.

These algebras are not only dual to each other, but they also have the property of being able to be represented on the same diagram. Any finite distributive lattice is, in effect, a "double Brouwerian algebra", i.e. both a Brouwer lattice and a Heyting lattice. As the product of a distributive lattice is a distributive lattice, we can consider the application of this structure to the different levels of realization of the Hegelian concept.

Before that, however, we can already note that the local complements (Heyting pseudo-complement and Brouwerian complement), perfectly formalize the Porphyrian mode of thought. With a concept being posited (A = Being), we have, in the finite, two forms of negation: a non-being according to the power ($A^* = B$) and a non-being according to alterity ($\neg A^* = \neg B = C$). Therefore, we can represent this in the following way:

Consider a double Brouwerian algebra $\mathcal{A}$ consisting of a set $P = \{A, B, C\}$ and four binary operations: $\wedge, \vee, \to$ and $-$, such that that $\mathcal{A}$ is a Brouwerian algebra under $[\wedge, \vee, -]$ and a pseudo-complementary algebra under $[\wedge, \vee, \to]$. The lattice of Figure 4.3 shows that we go from a position A to a position B by the pseudo-complementation $A \to B$. We then go from B to C by the Brouwerian complementation $\neg(A^*) = \neg B = C - B = C$.

Figure 4.3. *Distributive lattice on the Porphyrian three-element set*

The corresponding laws are indicated in the following matrices:

$\wedge$	B	A	C
B	B	B	B
A	B	A	A
C	B	A	C

$\vee$	B	A	C
B	B	A	C
A	B	B	C
C	C	C	C

$\rightarrow$	B	A	C
B	C	C	C
A	B	C	C
C	B	A	C

$-$	B	A	C
B	B	B	B
A	A	B	B
C	C	C	B

Now the transcendent negation of all the finite, in Porphyry, leads to the affirmation of the One, another name for God. This admits, as for Parmenides, a representation via projective geometry. This time, we must use PG[1,3], defined on the ring $(\mathbb{Z}/3\mathbb{Z}, +, \times)$, which is also a field and a vector space. The homogeneous coordinates of this projective line PG[1,3] will be taken in the ring $\mathbb{Z}/3\mathbb{Z}$ which has, as we know, the elements 1, 0 and -1, and whose addition and multiplication tables are:

$+$	0	1	-1
0	0	1	-1
1	1	-1	0
-1	-1	0	1

$\times$	0	1	-1
0	0	0	0
1	0	1	-1
-1	0	-1	1

As we have already shown (see Parrochia 1993a, pp. 225–226), the algebra of $\mathbb{Z}/3\mathbb{Z}$ is incompatible with the double Brouwerian algebra that we have defined. However, note that, since both 1 and -1 are units, the projective line has the four elements (1,0), (1,1), (1,-1), (0,1)[3]. An important observation is that there is no addition (comparable to that existing on PG$[n, q]$ when q is even) to be defined on this finite projective line[4]. However, we arrive at the following

3 The ring $\mathbb{Z}/3\mathbb{Z}$, which is also the Galois Field GF(3), has the elements 1, 0 and -1. Its projective line has the four elements $U_0 = (1, 0), U_1 = ([1, 1), U_2 = (0, 1), U_3 = (1, -1)$, since both 1 and -1 are units. Of course, the four points can be permuted in every way by the projectivity. Thus, the projective linear group PGL(2,3) $\backsimeq$ S_4, is just the permutation group or the elements of a four element set, and $|S_4| = p(2, 3) = 24$. Hence, PGL(2,3) acts sharply quadruply transitively on PG[1,3] and the four points form a superharmonic tetrad (see Hirschfeld 1979, p. 129).

4 The reason for that is obvious. It is a very well-known result of algebraic geometry that a projective curve with an algebraic group structure (as addition would be) must be an elliptic curve. Therefore, no combination allows us to pass from the points at finite distance $(1, 0), (1, 1), (1, -1)$ to the point at infinite $(0, 1)$.

representation of PG[1,3] that we can identify with the Porphyrian way of thinking (see Figure 4.4).

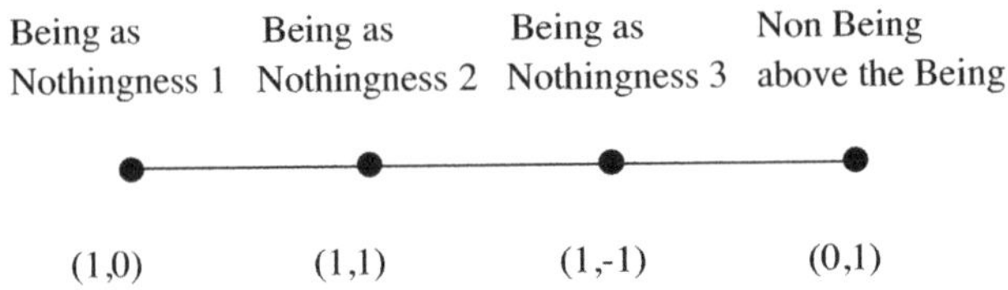

Figure 4.4. *The projective line PG[1,3] or Porphyrian way of thinking*

4.5. Quasi-Hegelian systems

Dubarle showed how the Hegelian way of thinking was a generalization of the Parmenidian one. In the type of explanation proposed, the binary Parmenidian basis gives rise to the finite projective geometry PG[1,2]. In the Hegelian case, it is replaced by a base defined on a set of four constants of structure, which then allows a finite projective geometry PG[2.2]: the seven-point projective plane.

In this new formalism, we just have to replace the set $P = \{A, B, C\}$ with the set $H = \{U, P, S\}^n$, associated with isomorphic matrices for the internal operations when $n = 1$. This is just a generalization of the Porphyrian mode of thought. Now, consider the set $H_2 = \{U, P, S\}^2$. It has $3^2 = 9$ constants of structure, which are the following:

$$H_2 = \{(U,U),(U,P),(U,S),(P,U),(P,P),(P,S),\\ (S,U),(S,P),(S,S)\},$$

where (U,U) means, as we know, "Universal of a universal", (U,P) "Particular of an universal", etc. Then, using the ring:

$$[\mathbb{Z}/3\mathbb{Z}]^2 = \{(0,0),(0,1),(0,-1),(1,0),(1,1),(1,-1),(-1,0),\\ (-1,1),(-1,-1)\}.$$

we can define a finite projective geometry PG[2,3][5], which is the 13-point projective plane, of which nine are at finite distance and four are in the infinite.

5 It seems that in his last writing on *L'Ontologie de Saint-Thomas*, Dubarle also compared the schematism of the last philosopher's system to the "algebraic-geometric coordination" associated with the projective plane PG[2,3] (see Dubarle 1996, p. 440; Courcier 2008).

With the same problem as previously to pass from finite to infinite, we get the following figure (see Figure 4.5), where the points are, as previously, labeled in homogeneous coordinates.

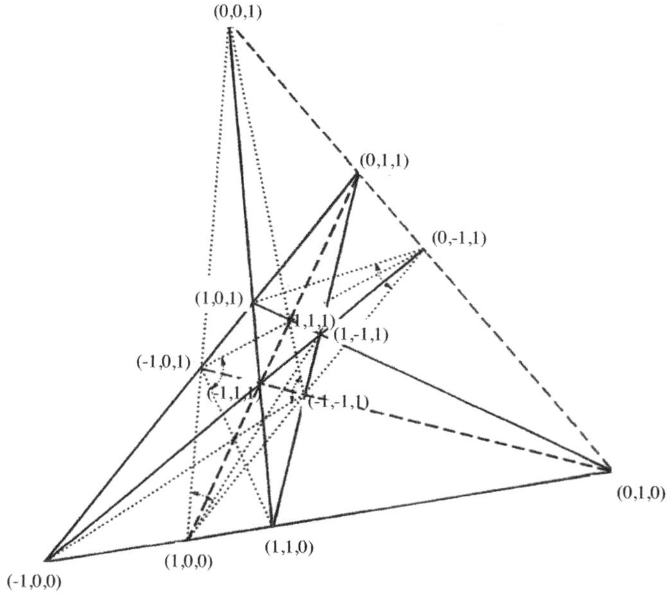

Figure 4.5. *The plane PG[2,3])*

This geometry PG[2,3] has 13 points and 13 lines. Each line is a projective line PG[1,3] with 3 + 1 = 4 points in it, and each point has four lines passing through it.

Relative to the Hegelian system, we can object that four points at infinity is one too many: three should have been enough. But this is to forget that, with each triplet at finite distance being linked to a point at infinity, it is normal that the triplet at infinity is itself linked to a point at infinity. It can be interpreted by noting that the infinite triplet $\{Universal, Particular, Singular\}$ is, in fact, the expression of the infinite Concept[6].

6 If this aspect was to prove too problematic, we could always use a possibility revealed by Hirschfeld (1979, p. 391): if a line and the four points on it are deleted from PG[2,3], there remains a (9,12) figure of nine points on 12 lines with three points on a line and four lines through a point. This is the arrangement of the nine inflexions of a non-singular cubic curve in PG(2,ℂ).

If we keep, for such a system, a hierarchical structure, it goes without saying that the subsequent realizations are 9^n in number and that the corresponding finite projective geometries are the PG[2n+1, 3] planes, which introduces a complexity that is quite remarkable, a way of super-Hegelianism. We will note, however, that the higher levels appear contained in the system, if we identify its primitive operations between the elements of the set $\{U, P, S\}$ to the multiple negation or transformation operations defined above.

We may now prolong the movement, and try to complete this beginning of classification.

Thinking cannot be done on the mode of identity or internal non-opposition. The mind cannot pose any entity, or say, its mnemic trace, except by opposing it to others in a certain fashion, which amounts to determining it in a particular way. On the contrary, it is legitimate to suppose that reason requires the whole of the thinkable, with the aim of reducing it to a set of generators, if possible finite. Henceforth, the thread that should make it possible to identify all the possible ways of thinking can only be found in the fundamental structures of algebra, to which the various types of known logic generally fall. If we then admit that finite traces are finite representations of elements possibly endowed with infinite semantic loads, then it remains to draw up a cross-tabulation of the set of finite structures and their associated projective geometries when they have them. But we can also admit that we must eventually consider infinite structures of which no finite realization can account. Gradually, we arrive at the idea of a general classification on the set of possible mathematical structures (see Pierce 1970; Parrochia 1990) as they are associated with particular modes of philosophical thought.

4.6. Philosophical thinking and finite projective geometry

Let us begin by giving some examples of classifications of finite structures, to which we can associate philosophically precise contents:

– the Parmenidian thought, as we have seen, is characterized by a simple Boolean algebra defined on a set $B = \{0, 1\}$ with two elements (the Being and the Nothingness, the one and the zero). It does, however, admit a representation in finite projective geometry;

– the Platonism, at the time of the *Republic* dialog (VI, 510 a sq), brought the whole of reality back to the set of four subsets $\{S, C, M, I\}$, where S denotes the images or simulacra, C is the set of things and the living, M is the set of mathematical objects and I is the set of Ideas. A structure of partial order is defined on this set of classes, which gives the greatest dignity to the Ideas, the smallest to the images, and puts mathematics and their main field of application (the living and real things) on an equal footing[7]. We have here a first type of structure, quite particular, which we can interpret again as Boolean – as it happens, a Boolean lattice isomorphic to $B^2 = \{0, 1\}^2$, with four elements – Ideas being the least upper bound, Image being the greatest lower bound, and mathematical objects and things or living beings standing as complementary terms;

– Hegelian thought, if we believe Dubarle's study, is drastically translated by a sort of relativization of the Parmenidian–Platonic Boolean structure to some regions of experience, the whole of reality, conceived of in a much more complex way. It then identifies with a product algebra $\{\Lambda, U, P, S\}^n$ of the algebra with four constants of structure, isomorphic to the Boolean algebra B^2. In addition, a projective geometry is here associated with it to express the transition from the finite to the infinite;

– we have previously shown that Porphyry thought, defined on a basis of three elements, could be explained in a finite projective geometry PG[1,3], and that the extension of such a system, defined on a basis of 3^2 elements, gave a projective geometry PG[2,3] in the form of the 13-point projective plane, even though, in the two cases, there is no algebra that makes it go from the finite to the infinite.

Assuming that there are many other possibilities, we can therefore draw up the table in Figure 4.6.

It remains to specify the content of empty boxes. The inventory of smaller finite fields already makes it possible to ask interesting questions.

We see that the Hegelian system in its most basic form (algebra with four constants of structure and the associated projective plane PG[2.2]) can be generalized in several ways. In particular, it is important to find other correspondences between fields or finite algebras and projective spaces. From then on:

– we remain in the plane, and we have the succession of possible projective spaces PG[2, n];

– or we go to spaces of higher dimension, for example, PG[3,2], which amounts to defining a set of 15 planes, each having seven points as previously

7 It is quite easy to prove that the famous line divided by Plato in a certain ratio k, then further subdivided in the same ratio, has central segments that are equal (see Parrochia 2018, pp. 17–20).

in PG[2,2]. PG[2,2] consisted of seven lines with three points each. PG[3,2] is formed of 15 planes comprising seven lines and seven points each. Everything happens as if the first projection (Fano's configuration) should itself be considered as one of the many finite traces of a much richer space since it has one more dimension.

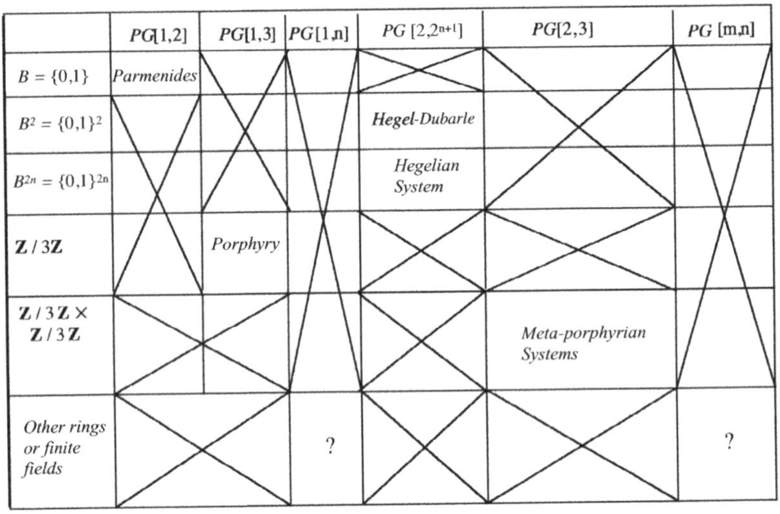

Figure 4.6. *Hegelian, Porphyrian and Meta-Porphyrian systems*

Beyond that, of course, we can list all the possibilities of thought associated with a representation of the infinite in the finite. However, we know that projective planes do not always exist. Thus, if the order n is a power of a prime, i.e. $n = p^a$ for $a \geq 1$, it is conjectured that only projective planes of order p^a can exist, i.e. projective planes of order 2, 3, 4, 5, 7, 8, 9, 11, 13, 16, ... But there would be no projective planes of order 6, 10, 12, 14, 15, ... Therefore, mathematics put some limits to philosophical imagination, even though the number of possible isomorphic structures can sometimes be very large (see Moorhouse (2006, 2020) and Table 4.2).

4.7. Other algebras for philosophical thinking

If fields, rings and finite groups are structures of choice for the expression of thought, less rich structures such as partial-order structures (lattices, semilattices) or their algebraic equivalents (semigroups, semirings, modules) can already be used, in some cases, the expression of value hierarchies. As we have seen, a second generalization of Dubarle's formalization then concerns non-classical logics, those which, in their algebraic expression, lead to particular lattices (relatively pseudo-complemented, pseudo-complemented, contrapositional, etc.) and to their numerous isomorphic algebras: Heyting–Brouwer algebras, Post algebras, etc. (see Rasiowa 1974).

n	Number of planes of order n	n	Number of planes of order n
2	1	16	≥ 22
3	1	17	≥ 1
4	1	19	≥ 1
5	1	23	≥ 1
7	1	25	≥ 193
8	1	27	≥ 13
9	4	29	≥ 1
11	≥ 1	...	...
13	≥ 1	49	> 500,000

Table 4.2. *Number of planes up to isomorphism (i.e. collineations)*

It will be noted that the paraconsistent logics give rise to algebraic formalizations and special lattices that could have an interest in epistemology. Although we are generally limited in not considering theories in relation to each other, the current situation in science is to face theories that do not allow (or not completely) a connection to each other. A coherent representation of such a situation is made possible by paraconsistent logic (see Arruda et al. 1980).

We could also consider extending the expressive possibilities of algebra in philosophical matters, to non-associative algebras, which play a role in genetics and which could possibly also serve to explain certain modes of generating ideas and ideal numbers in the last Platonism.

4.8. Models derived from geometry and algebraic geometry

Classical surfaces already constitute philosophically important structures, in relation to both mechanics and optics. Having become differential, geometry allows many expressions, not only of the spatio-temporal continuum, but also of abstract spaces parameterizing the evolution of physical entities and people.

A quick classification will be proposed here, in connection with the uses (metaphorical or not) that authors have been able to propose in the story.

Call P_2 the projective plane defined by the union of a disk D and a Moebius band M of intersection $S = D \cap M$, homeomorphic to a circle.

Let U_q be the connected iterated sum of p copies of the projective plane $P_2 = U_1$.

Now let us call T_1 the torus. For $n \geq 1$, V_n will be the space obtained by removing a small disk of T_n and T_{n+1} will be obtained by gluing V_n and V_1 along their edges (see Figure 4.6). We posit that T_0 is the sphere S_2 and V_0 is the disk D_2.

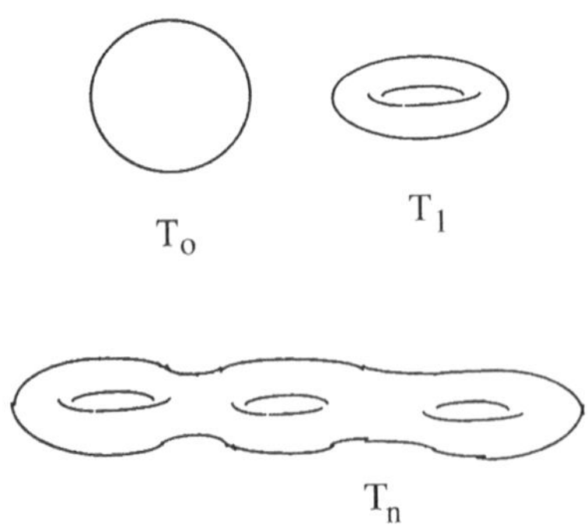

Figure 4.7. *Classification of surfaces*

Given these definitions, it can be shown that any connected surface is homeomorphic to either a space $U_q(q \geq 0)$ or a space $T_p(p \geq 0)$.

There are therefore only two possible classes of geometric manifolds of dimension 2: one homeomorphic to the projective plane or its iterations, and the other homeomorphic to the torus or its iterations.

It is, in fact, the study of the Morse functions on a related compact differentiable manifold M that allows this heuristic classification that sends M on one of the spaces U_q or T_p. Conversely, we can show that each of these spaces is homeomorphic to a differentiable manifold, so that we have a complete enumeration, with homeomorphism, of all the related compact differentiable surfaces. It seems that this nice result is peculiar to dimension 2 and does not have an analogue for the varieties of dimension $p \geq 3$ (result of insolubility of Markov (see Markov 1958; Gramain 1971, pp. 39–40).

The interest of a classification of geometric varieties is obviously that philosophy, especially in the 17th century, is essentially a philosophy of geometers (Descartes, Pascal, Huygens, Spinoza). The essence of the Cartesian representation of the world refers to a linear structure, but Huygens, and perhaps, in his wake, Spinoza, think in terms of curves (circle, cycloides), even of developable surfaces (nature as a geometric entity). Most of the time, for these authors, all of these surfaces remain surfaces and do not explicitly serve to model the world or parts of it. It will not be the same in modernity when Lacan will use the metaphor of the Moebius strip to describe the structure of the unconscious or when, assuming loops and supermarkings, Desanti will propose, as a metaphor for the real, a Turing supermachine actually quite unorthodox (see Desanti 1980).

If the introduction of recursive configurations in Pascal (Latin squares, Pascal's triangle, etc.) does not go much beyond a very elementary algebra, the manipulation of the Spinozist infinites would require, to be well conducted, Riemannian notions, which will only intervene in the 19th century.

Hence, philosophical geometries are often so disappointing. Hegel will have nice game to defend his conceptual approach possessing a true internal richness.

But algebraic geometry has progressed a great deal since Descartes, and we will probably wait a long time – for philosophers today encounter enormous difficulties in approaching mathematical thought – for the philosopher of contemporary algebraic geometry. In this panorama of possible pure philosophies that we are sketching, we would like to say one more thing: there is no doubt that it would be necessary to start from the extension of the basic field to complex numbers, to study the methods at work on algebraically closed fields, and to be interested in what makes the generality of the results of algebraic geometry since the 19th century, namely, the consideration of projective algebraic varieties, parts of a projective space $P_n(\mathbb{C})$. For 50 years, the central problem of algebraic geometry has been to generalize the properties of classical varieties defined on $\mathbb{C}$ to varieties defined on any field, an extension that has brought particular new methods. The idea of a possible contraction of the curves and algebraic varieties led to the project of a classification of their minimal models. This is almost complete for surfaces, and it seems that some of these results can be transposed to larger varieties (see Dieudonné 1977, pp. 140–141). Assuming that a philosopher is interested in these questions, they would find good examples of invariants through the theory of classical invariants, created in the 19th century with a view to a classification of algebraic subvarieties of a projective space, which has become part of the theory of linear representations of groups, in the literature (see Hilbert 1993). Going back and forth from Cayley and Hilbert to Dieudonné (see Dieudonné 1974) would obviously be of the greatest interest, with a view to producing a true philosophical method worthy of today's mathematical thought – an approach that Bachelard himself already advocated, if not anticipated[8].

4.9. Conclusion

One of the major problems of contemporary philosophy is to arrive at producing concepts that are appropriate to the complex intellectual universe in which we live, and of which it is not certain that it is redescribable by means of concepts: these have, indeed, the defect of their origin. Born to solve specific problems, in epistemological contexts much simpler than the one we are confronted today, it is not surprising that they seem very insufficient to us to organize our experience and account in a relevant and creative way. The thread of mathematical structures commits us to generalize them, and at the same time to produce extraordinary extensions of the concept of philosophy. At the beginning of the 21th century, after a hundred years which have witnessed the development of so many beautiful formal constructions, it is really surprising to see the emergence of a thoughtful philosophical bourbachism that operates, with reference to conceptual forms, the generalizations led by a powerful mathematical way of thinking.

Is the perspective we are advocating a culpable transgression? Is our approach fundamentally different from the one Plato once attempted to achieve by transposing in philosophy Pythagorean speculations on continuous fractions, the dimidiation of unity, and the rationalization of the *aloga*? Ideas that he came to define with the help of convergent dichotomous processes, successive series of conceptual divisions, making them appear mixed, were a first step – a first step that we only sought to extend. One day, perhaps, the conceptual transpositions that we suggest will appear as banal and natural to the philosophers of the new times as the Platonic classifications can be for us today. In any case, it is our hope that an immense and unexplored universe would then open up to future philosophy.

8 See the very nice use of conformal group, birational geometry and Cremona geometry in Bachelard (1949); also see our commentary in Parrochia (1997) and on disciples or followers of Cremona like Enriques (see Castellana 2019).

5

Philosophical Transformations

Let us pose the problem in all its complexity: how can we think rationally about (and maybe formally) the relation of philosophy to its history? The characteristic of a philosophical doctrine, in fact, is not only used to pose as a truth among others. It tends, more or less, to present itself as *The Truth*. The peculiarity of each philosophical doctrine is to wish to pose itself as *The Philosophy*, i.e. the only possible philosophy in the world. At the same time, and correlatively, each philosophical doctrine appears in the course of history and it is only *one* philosophy *among others*. The great scandal for all philosophy is the plurality of philosophies. How can we solve this problem?

All philosophy is caught up in a paradox: to exist, it needs other philosophies, especially the philosophies that preceded or are contemporary to it, philosophies of which it discusses theses and in relation to where it is positioned; and at the same time, all philosophy seeks to abolish the scandal of the plurality of philosophies, it seeks to put an end to the history of philosophy, to take the place of all the other philosophies which preceded it in the course of time.

5.1. The paradox of a metasystem

One solution could then be to think systematically about the plurality of philosophies. It would amount to considering the construction of a kind of "metasystem" or "system of all systems".

1) firstly, it is necessary to question the very notion of systematicity: in order to know if a system of all systems is possible, we must first know what a system is and whether a (philosophical) system is possible. We will see that a number of philosophers have rejected the idea of a system. In the name of what? We will

have to determine it. But we will show that these critics of the idea of system strengthen it, rather than challenge it, because systematicity has many figures and we do not get rid of them as easily as we say;

2) if we accept the idea of a system, how is it then necessary to conceive a systematic thought of the plurality of doctrines? Is it on a classificatory and a-historical mode? Is it, on the contrary, in a Hegelian mode? We will return back to back these two solutions as unsatisfactory ones because they both misunderstand the reality of the doctrines;

3) how not to misunderstand the reality of doctrines and how, at the same time, to strive to think systematically of their links? Here, we seem caught in a double trap: either we build a new philosophy, and in this case, it crushes or distorts the others: we then loose these last; or it is limited to making history of philosophy: it deploys well, at that time, attention and fidelity to the doctrines, but we lose the rationality of the whole.

After reflection, we see that there are only two solutions: Martial Gueroult, in a perspective close to Kant, tried to found, with what he called "Dianoematique" (see Gueroult 1979), a systematic doctrine, but who strives to allow the existence of others. We will show that his solution is not tenable. The other solution, which is the one we advocate, is to make a kind of science of philosophy: this science will always be partial, but it will try to develop invariants.

A) Is the idea of systematicity questionable, even, strictly speaking, unthinkable and why? What are the arguments that can be made against the system, and which are therefore, a fortiori, against the system of systems, the metasystem envisaged?

There are at least four ways to question philosophical systematicity as follows:

1) we could not isolate the system, so we cannot constitute a closed set, so there is not, strictly speaking, a system. Just as a logic of the "supplement" characterizes any encyclopedia, a logic of the *par-ergon* would anime any work, all "erga". We will recognize here a Derrida objection: the conceptual arrangement of the philosopher dissolves in the very systematicity of language, which passes through it from one side to the other and flees on the timeline;

2) assuming that we can define the system as a subsystem of language, must we not then be related, in order to be fully understood, to the material conditions that determine it: the politico-economic frame (materialist historiography from Marx to Lucien Goldman), the state of sciences and techniques (M. Serres), the very lived experience of the philosopher (S. Freud, G. Bachelard, Y. Brès), which can, to a certain extent, influence his thought, etc.?

3) let us admit that the system escapes. Let us consider it for itself: does it have, for all that, the coherence that historians such as Martial Gueroult lends to it, which is that of a static monument? When we analyze it closely, including using quantitative studies, lexicographical indexing and taxonomic inventories, we sometimes make appear disjunctions, flaws, which can ultimately break it. Or, at the very least, we can sometimes also spot in it a periodicity. André Robinet, in his many computer works on Spinoza, Malebranche, Leibniz or Rousseau, did not fail to arouse such objections;

4) suppose, however, that, in spite of its faults, the structure is saved. It is only the organizational structure, i.e. a quasi-formal framework. But what about argumentation? Must not every philosophy, to the very extent that it must communicate itself, target an audience and consequently, adapt its language to it? This supposes that the philosopher must make use of rhetorical methods. But, what kind of story do you have to make? That of the systems, or that of the procedures of argumentation, in other words rhetorical places or toposes, modes of reasoning, elocution or composition of speeches? Once again, the system is dissolved in its conditions of possibilities. Chaim Perelman, in *Rhétorique et Philosophie*, or in his *Traité de l'Argumentation*, inaugurated another way of apprehending philosophical doctrines, that moreover Anglo-Saxon philosophy, although independently, practices more willingly than structuralist historiography. We then wonder how a philosopher reasons rather than what he says exactly.

These four objections, that we unfortunately cannot develop, seem to make the notion of system multiply. But a system of all systems remains a system. Therefore, we find at its level all the problems that can be raised at the level of its "subsystems", if we are allowed this expression.

For all these reasons, many philosophers have refused, not only the idea of a systematicity, but a fortiori, that monstrous systematicity that would be the system of all systems, utopia not only impossible in fact, but also in law particularly presumptuous, since it intends to put an end to the history of philosophy: is this final solution, which wants the death and the extinction of other systems, then not the most dangerous of lures? Is this not pure madness?

We would like to prove here that this is not the case, starting by making the following observation: is the worst system not the one that exists but that we do not notice? Therefore, by refusing to assume the existence of systematicity in thought, is it not to leave thought to an implicit, unconscious systematicity, and hence, all the more alienating that we do not control neither the effects nor the limits?

Therefore, we have no choice: we have to bet on the system, the consciousness of a culture, even though the system is not everything: just as we must strive to make ourselves aware of ourselves and its acts, even though this effort will never eliminate the unconscious part which is in itself.

For all these reasons, we must therefore assume the system possible, and even necessary.

B) Assuming that we accept the idea of a system, how should we conceive a systematic thought of the plurality of doctrines? The concept that is in question here is that of a "logic of philosophy". This idea of a logic of philosophy was born and presented at first in a metaphysical form with Hegel. Then, it took different figures that will be analyzed later. Let us first study the Hegelian case.

The Hegelian system, let us recall, presents itself in the form of an encyclopedia of which the first moment is logic, the second is the philosophy of nature and the third is the philosophy of mind. The Hegelian system must therefore present in a logical way, i.e. articulated according to the three moments of the dialectic, all the spiritual realities, including the absolute spirit, i.e. art, the religion and philosophy, which have a historical future. In separate books, Hegel therefore tried to systematize these domains, in other words to find the categories constructed in the "Science of Logic" (first moment of the *Encyclopedia*). In the *Lessons on the History of Philosophy*, Hegel strives to present a kind of logic of this history in the form of a series of successive dialectics (of the universal, particular, singular type), the moment of the particular being the driving moment, the properly dialectical moment or moment of contradiction. What is the Hegelian method and what problems does it pose?

In the introduction to the *Lessons on the History of Philosophy*, where Hegel explains his method, we note that he exposes a certain number of licenses that he says he must take with the concrete unfolding of events:

1) he will not insist on the contemporary political events of the philosophies, nor on the biography of the philosophers;

2) he will only consider the philosophies "whose principle gave an impulse" to the history of thought – in other words, the principal philosophers, the inventors and not the diffusers: "any system, indeed, counted a number of teachers and followers; we might, therefore, point out a host of names of the men who, in part, have had great merits as professors of philosophy; however, we must also neglect the mode of propagation of a philosophy";

3) finally, the philosopher will consider above all the principles of the systems, to the exclusion of their concrete applications.

In other words, a triple filter is supposed to recapture pure action, "sanctioned" philosophical thought (to use a Bachelardian expression): drowned in political and event history, this thought is first liberated; then, the second moment, eliminates the repeaters; finally, the system is projected on its principles – its key schemas.

From the abstract to the concrete, and according to ternary dialectical concatenations, the systems will then align, in theory, in the order of time.

Does Hegel fully realize this vast program? When we look at the results, we see that its logic globally respects the chronological order; at least as regards the principal doctrines (thus Plato comes before Aristotle, Descartes before Spinoza, etc.). However, for some periods, permutations will be noted. In medieval philosophy, for example, the scholastic moment unfolds as follows: Peter Lombard (mid-12th century), Thomas Aquinas (1224–1274), John Duns Scot (1266–1308), come before – in this order – Alexander of Hales (died in 1245), Albert the Great (1193 or 1205–1280) and Gautier de Mortagne (1174).

The license with the historical unfolding is justified by the fact that the first authors are systematicians of the new scholastic methodology, while the second ones are mainly commentators of the writings of Aristotle: in other words, they synthesize the spirit of the old philosophy and that of the new. They are therefore authors who must appear locally in a third moment, the moment of synthesis. Hence, they come after others.

Therefore, it seems that in a half-century period, the philosopher gives himself the possibility of organizing the period as required by the fundamental principles of dialectic and not the servile respect of chronology.

Another problem is knowing when philosophy really begins. There, either by a desire for symmetry with respect to religious or even general history (which is obviously older than philosophy), or, rather, for the sake of not excluding entirely the period corresponding to pre-philosophical representations (the philosophemes), Hegel, after having said that philosophy began only in the West (where the spirit, entering into itself, posits itself as free and becomes free for itself), reintroduce considerations on the "oriental philosophy". This is, indeed, the true "root" of the tree of systems, which springs up from the ground on Greek soil.

Therefore, let us draw the conclusions of this brief analysis. In this domain of the history of philosophy, characterized by a certain complexity, Hegel simplifies and reorders reality by means of a logic that is at least partly external to it. Hence alterations, not only of the chronology but also of the interpretation.

These limits have thus cast a certain discredit on the idea of a historical logic of philosophy. But other projects, later, appeared.

Martial Gueroult, who accounts for this in his article published in a French encyclopedia on logic, architectonic and constituent structures of philosophical systems (see Gueroult 1957), summarizes the possible positions in the following way: the logic of philosophy, when it is not, as in Hegel, of an historico-metaphysical type, can be conceived in several ways:

1) as a problem of methodology (Benedetto Croce, an Italian philosopher of the beginning of the century, and still quite close to Hegel, considered philosophy in this way);

2) as the elaboration of a theory of categories peculiar to the philosophical region (a German philosopher of the early 20th century, E. Lask (1911), rather conceived this logic in this way);

3) as the study of the genesis of philosophical concepts (that is how the French philosopher Leon Brunschwicg, in his studies on causality or the formation of mathematical thought, included philosophy). This form of continual conceptual logic has given way, notably to Bachelard and Canguilhem, to a discontinuous perspective;

4) finally, the idea of a logic of philosophy can be conceived as the determination of a certain number of essences typical of systems which must be disengaged in their abstract purity. Aristotle and Carnéade can be considered as the initiators of this last perspective: in particular, Aristotle interprets all the systems that preceded him through his theory of causality, each system illustrating one or other of the four typical causes (formal, material, instrumental or motor, final).

But this perspective, inaugurated by Aristotle and Carnéade, of systematically thinking philosophy through a logic that identifies the essences of systems, can itself take on various aspects:

– in the 19th century, it was Renouvier who carried this project further with his sketch of a systematic classification of philosophical doctrines, based on a series of six alternatives: The thing, the Idea; the Infinite, the Finite; evolution, creation; Necessity, Liberty; Happiness, Duty; Evidence, Belief. Against the eclecticism of Victor Cousin, who claimed to reconcile all systems, Renouvier made manifest irreducible oppositions and resulted in a classification of neokantian background, neo-criticist inspiration;

– there are other versions of this systems logic. For example, Wilhem Dilthey, in *Die Typen der Weltanschauung* ("The types of the world view") (1911) presents a more psychologic version of this logic: he shows in particular that we cannot understand the logic of the sequence of philosophical systems without comparing philosophy with art and religion, which are also manifestations of spiritual life. Now our experience is always given as a world in which, as in the Spinozist imagination, everything is enveloped and in close correspondence. It is therefore necessary to link the systems to the great psychic groups which are at the origin of them and which correspond to certain degrees of differentiation of this imagination. Dilthey's work is unfortunately unfinished and has not resulted in a fine classification of history;

– we could highlight (as Martial Gueroult does) still other attempts: psychologist-logicist (with Hoffmann), psychologist-metaphysics (with Jaspers), logicist-ontological (with Gilson) or even logicist-transcendental (with Karl Mannheim).

The latter is the deepest. He has particularly endeavored to explain in philosophical systems the structure of the theory of knowledge, in order to index them by this means. Let us quickly give an overview of his method. According to Mannheim (Mannheim 1964), knowledge is broken down into different regions such as psychology, logic, ontology, etc. For each of these regions, there are a couple of key philosophical concepts related to it. For logic, it is the couple form/content; for the ontology, it is the couple substance/accident, etc.

In the field of the theory of knowledge, it is the subject–object pair that is fundamental. But the theory of knowledge, to be systematized, must have recourse to other domains such as logic, psychology or ontology, and it is conceived differently according to whether it gives the primacy to one or the other of these areas.

For example, if the primacy is given to the ontology, then the theory of knowledge will suppose the idea that the same ontic matter constitutes the subject and the object, and that the relation between the two is originally given. This is the case, as we know, in Leibniz, where we start from this deep identity to understand both the relation of the subject to knowledge and the relation of knowledge to the object: the monadological interexpression is based on all that. This proximity of matter and spirit makes Leibniz say, for example, that matter is a momentary moment, a momentary spirit.

If, on the contrary, the primacy is given to logic, then, in this case, we start from the relation of knowledge, as objective knowledge, to the object, from which we can seize the relations of the subject to the object and subject to knowledge (Marburg School).

If the primacy is given to psychology, then we start from the relation of the subject to knowledge (as consciousness), to grasp from there the two other relations (subject to object and knowledge to object) – which is the choice of classical philosophy, namely, Descartes.

If we recapitulate, we see that we have a theory of knowledge that stages subject, object and knowledge itself. And we also have the choice between three major orientations: ontological, logical and psychological. But we must also see that the form of knowledge will vary according to its point of departure: either the subject or the object or something beyond both:

– if it is the subject, the knowledge will take the form of a reproduction of the object by the subject (thus a copy: this is the case with Kant);

– if it is the object, knowledge will take the form of a production (it will manifest a spontaneity of being, as in Spinoza's philosophy);

– if it is a being beyond both, then the knowledge takes the form of a preformation or a pre-established harmony (as in Leibniz).

From this double combinatorics, a large number of subtypes can be generated. We will just give an example.

Suppose that in the theory of knowledge, logic is dominant. The subject–object relation is therefore articulated according to the fundamental form–content couple, which is that of logic.

Then, sometimes the subject will be put into form and the content is then deduced from the object; sometimes it is the object that will be in form and content that will be subjective.

Then, this form–content relationship can be thought of three or four different ways:

– or the content is reduced to form (Marburg school);

– or the form is reduced to the content (logical realism of Lask);

– or form and content coincide in a higher metaphysical principle (this is metaphysical realism);

– or there is a radical split of form and content (as in Kantian criticism).

We could take ontological and psychological first types and make similar reconstructions.

Therefore, what makes these philosophies of philosophy, in spite of their subtlety, do not win the adhesion? Certainly, there is a seduction of these logics: they make it possible to present the range of all the structural possibilities of philosophy from a single point of view. As a result, not only can the relationships between the various philosophies be clearly grasped, but also the rational classification makes it possible to clarify in advance the ways of their concrete analysis. These logics have advantages. And yet, according to Gueroult, we must reject them, in the name of two arguments:

1) these logics are only systems of philosophy among others;

2) far from describing systems from the inside and being attentive to their particular problems, they tend to relate them to pre-established *types* of which they would at best only constitute more or less successful examples. Systems are then interpreted as deviations from a norm imposed on them from the outside.

Martial Gueroult therefore strongly criticizes this perspective and substitutes for it the idea of a plurality of logics of philosophy, any system having its own.

Does this mean that this plurality is not then thought of systematically? Of course not.

As soon as it is thought philosophically, this plurality is thought systematically, but it is thought in the context of a system that does not intend to mislead others but let them exist side by side. This is what Martial Gueroult explains in a book called *Philosophy of the History of Philosophy*, in the volume entitled "Dianoematique" (Dianoematics) (see Gueroult 1979).

Dianoematics is the discipline that makes it possible to consider thought transversally, considering that philosophical systems are the true ideas of thought, in other words, both the essential possibilities of thought and consequent thoughts, i.e. thoughts whose development has been logically pursued to completion.

Why are these system-ideas fundamental? Because, without them, the reality would be purely indeterminate. What allows us to define reality is philosophy, it is the judgment by which each philosopher decides what reality should be by posing his own doctrine as a *thesis* on this reality (*thetic judgment*).

In other words, the doctrine presented by Martial Gueroult is a kind of philosophical idealism which posits that the actual reality is not the indeterminate common real, but the philosophical reals each time determined by the philosophers in a particular way. Each of these system-ideas determine in its own way what reality must be, including its scientific, political, moral and other consequences.

Each system-idea, therefore, somehow explains a kind of timeless essence of thought. In other words, the indefinite collection of these ideas constitutes the totality of what is thinkable.

And thus, we see that the doctrine by Martial Gueroult – this radical idealism – generalizes here, at the level of philosophies, a structure that Spinoza invented and that he called the infinite understanding of God, in which hierarchizes more and more ideas, from the poorest to the richest. In Gueroult, the ideas of this understanding are philosophical system-ideas (each of which, as we have seen, having its own logic and architectonics) and these system-ideas, which are in a sense the "souls" of their epochs, are ordered in this infinite superstructure, this supermode of thought which constitutes almost for him a proof of the existence of God.

Therefore, we believe that this doctrine, which is therefore a kind of meta-doctrine, meets a paradox (moreover, seen by Martial Gueroult himself) and that we commented in the first chapter of one of our book (see Parrochia 1993a). This paradox is the following: every philosophical system is defined by two pretensions:

1) to be exclusively valid in an absolute sense;

2) to pose as being, in its content, a true image of the original (or, on the contrary, to demonstrate the impossibility of the image in general).

Of course, these two conditions must be reduced to mere pretensions. Martial Gueroult shows that, for the other systems to exist, it is necessary to limit these claims: any system believes to be absolutely worthy, believes to be a genuine image of the original or believes to demonstrate the impossibility of such an image. Well, all this only applies to the system in question, not in absolute terms. But the problem is that Martial Gueroult wants his metasystem to be a system.

It follows the idea that nothing forbids the application to the metasystem of all the properties that are valid for the systems which are in a sense its "elements": distinction between the concrete real and the philosophical real, claim to be absolute, condition of limitation, restriction of the claim to absolute validity to validity in its own sphere, etc.

For Martial Gueroult, this limitation should make it possible to prevent his own doctrine – the radical idealism – from leading to the falsity of all the other systems, thus contradicting his initial project.

In fact, if he had said that his doctrine was the truth of history, it might have been objected that he made a logic of philosophy, i.e. a blind system, which pretends to base the validity of others and, in fact, presents them in a biased way. Moreover, it could even be accused of contradiction since he said that any system must have its claim to absolute validity, and it would not have limited its own claim to this pretension.

1) On the one hand, the problem is that, by applying to the metasystem the same criteria as those of the system, Martial Gueroult settles into a paradox, formally comparable to that of the set of all sets, or to that of all sets which do not include themselves as an element:

- if its system is a system like any other, it belongs to the set of the systems. Therefore, it cannot be the system of all systems, since it is part of the whole;

- if his metasystem is different from the others, then it has been able to total the other systems, but it is not part of the whole, so it is not a system.

As Gueroult claims both to have made the system of all systems and to have also constructed a system comparable to any other, he is in a paradox. In this case, the problem is easy to solve: it is enough to admit that the collection of all the systems is not a set, or let us say, not a system, in any case not a system comparable to the others, but rather a "metasystem".

2) On the other hand (paradox of the system of all systems which are not elements of themselves), the contradiction results in fact from a situation well known to the mathematicians of set theory since Frege and Russell. It is the fact that one defined an object (in this case, a philosophical metasystem) by a property that is too strong (the claim of being a system like the others).

We also see, therefore, the possible solution. It is enough, this time, to define philosophical systematicity by a weaker property, which would not lead to the same consequences and maintain a heuristic distinction between the systems and the metasystem, without it automatically having to put an end to the philosophy or claim any kind of normative privilege over others.

This final solution seems to arise from the very text by Martial Gueroult when he defines the specificity of his own approach, i.e. of his own *thetic judgment*: "It is no longer a question here of creating a new metaphysics, but of to accept, with their own characteristics, those which history offers us, without wanting to maim them in advance, or to deny them in a total or partial way", Gueroult writes.

This statement in fact, if we wanted to be rigorous, is a meta-metastatement. This is a statement where Gueroult, logically, talks about his metasystem, so settles into a meta-meta-system.

But consider this phrase from a "naive" point of view: what does it mean? It means that it did not act for him to create an unprecedented form of metaphysics but to use existing forms to base their own possibility.

Let us accept this formula. Such an assertion implies, however, that "ordinary" philosophical work – metaphysical creation – is therefore always possible, although the author has personally decided that, as far as he is concerned, such a creation would not be his doing. If such a proposition does not appear in this context, it appears explicitly in paragraph 59 of his book, where the potential infinity of the becoming of philosophy is affirmed.

Then, we will ask the following question: how can metaphysics still be possible, even though philosophy has become, in the system of Martial Gueroult (but his work is only a symptom) conscious of itself?

The way in which other dianoematic systems can be created is easily conceived. But what about other metaphysics? Each philosophy "can only live by the death or absorption of the other", writes the author. At the same time, and by his task as a historian, he puts an end to this "immortal struggle", not only in a work that claims to have found their common possibility but also in a daily task in which he has tried to substitute the mutual ignorance of the authors' careful attention to the detail of their productions.

How can we think that the systems to come, if they can still be built, will not be affected in any way and resume, as in the past and in total somnambulism, their immemorial struggle?

In other words, what happens to the system when its fallibility is attested, when its claim to total jurisdiction, self-sufficiency, "surpassing" and excluding others are reduced to the idea of pure pretension?

We must ask what this situation means. Is it only an individual decision that one day led Martial Gueroult to estimate that, from then on, it was no longer a system that must be made but a metasystem?

Does this not mean that, to make a system of philosophy like any other, in other words, to make a system that results in the death of others?

But from then on, what Martial Gueroult sought to do was no longer exactly philosophical, and his metasystem is no longer exactly a system comparable to those of Plato, Kant, Hegel or Spinoza. What he did, what he sought to do, was to explore the possibility of philosophy objectively.

The questions that we would like to ask, in the end, would therefore be reduced to the following three:

1) can we, and to what extent, constitute an objective science of philosophy?

2) what limits does such an enterprise meet?

3) if such a project cannot escape philosophy, what is its least "expensive" philosophy? And, from this point of view, is Martial Gueroult's methodology the most appropriate? Is its underlying philosophy the least expensive?

Therefore, to try to escape what might be called the "Gueroult paradox"[1], we can develop a more modest project. It would be a question of describing what, in philosophical systems can be identified in an absolutely objective way by means of precise structures: for example, conceptual relations, which will take a hierarchical (arborescent) or graphic form (networks, graphs), so that the metasystem, which articulates all these structures, can be very strongly differentiated from these very structures. This avoids the paradoxes, while maintaining the objectivity of the approach and respecting the presuppositions of each philosopher.

As a toy-model, we can introduce such an attempt here.

It is known that philosophical doctrines, in their fundamental choices, cut reality in different ways, and that, from one doctrine to another, these cuts, which are in reality partitions or classifications, are very different. But it can happen that, on the same set of phenomena, these divisions can also differ[2].

---

1 This looks like also the famous "paradox of the preface", where an author sometimes states theses in contradiction with those which you will find in the chapters of his book, or, at the very least, which are weakenings of the statements of the book itself (see Makinson 1965).

2 Quine shows that the same is true within languages. In Quine (1969), he takes the example of Japanese classifiers. Attached to a number (*b*), a classifier (*a*) applied to an individual term (*c*) such as "beef" numerically modifies this term which remains an individual term. Nevertheless, if (*a*), instead of being attached to (*b*), is rather attached to (*c*), then the term (*c*) becomes a mass term (in the example quoted, instead of "ox", we will have the meaning of "cattle"). Quine thus justifies his thesis of the inscrutability of reference, since we do not know whether to interpret the sequence of words which includes a classifier, a number and a term according to the partition (*ab,c*) or according to the partition (*a,bc*). As this is only an example, we do not enter into subtle discussions of Quine's thesis, especially by Japanese authors (on this, see Takashi 1998). In philosophy, we could cite the case of the structure of Hegelian logic: according to Hegel, it is appropriate to attach the part concerning being (*a*) to the part concerning essence (*b*), thus constituting the *objective logic*, which opposes the doctrine or concept (*c*) or *subjective logic*. On the contrary, we could think – like Spinoza – that the domain of essence (*b*) or concept (*c*), which constitutes the domain of reflexivity (*bc*), are a kind of mirror reality that exists in front of (*a*) as the infinite mode of thought in front of the infinite mode of extension.

Anyway, suppose therefore, for the purposes of the case, doctrines which, locally, faced with three realities. Therefore, we start with a set $E = \{a, b, c\}$, and we assume we have the choice between the following distributions: $(ab, c), (ac, b), (bc, a)$[3]. Suppose now that, as in the case of Platonic dichotomous processes, starting from the total class of phenomena (in this case, (abc)), we want, by a kind of filtration process, to isolate one of them, i.e. for example, to pass from (abc) to (ab) and then, to (c). Now, if we want to enumerate all possible cases of this kind, we must first construct the lattice of partitions L on the set $E = \{a, b, c\}$, then, from there, the lattice of its sublattices.

The lattice of partitions on a non-empty finite set has well-known properties: he is semi-modular, graded, of finite height and satisfies the Jordan–Dedekind condition (see Parrochia 1991, p. 61). In the above case, where the starting set has three elements, it takes the following form (see Figure 5.1).

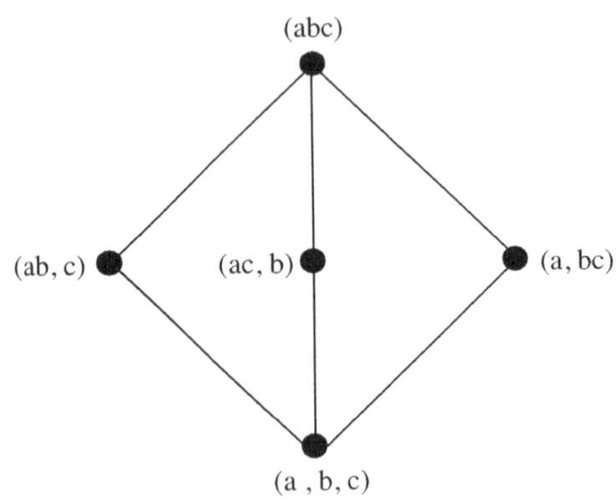

Figure 5.1. *The lattice L of partitions of $E = \{a, b, c\}$*

Now, for pure convenience, we will replace (abc) by 1, (a, bc) by 2, (a, b, c) by 3, (ab, c) by 4 and (ac, b) by 5. The lattice of sublattices LS of L takes the form of Figure 5.2.

Let us now introduce the following definitions:

DEFINITION 5.1.– *A subset I of a partially ordered set $(P, \leq)$ is an ideal, if the following conditions hold:*

1) I is non-empty;

3 We adopt the usual notations in classification theory, namely that $\{\{ab\}, \{c\}\}$ is simplified to (ab, c).

2) for every $x \in I$ and $y \in P$, $y \leq x$ implies that y is in I (I is a lower set);

3) for every $x, y \in I$, there is some element z in I, such that $x \leq z$ and $y \leq z$ (I is a directed set).

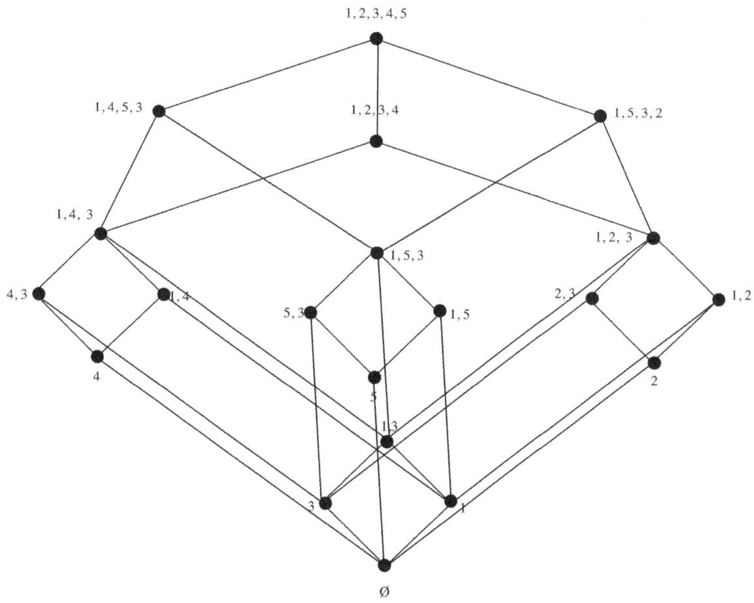

Figure 5.2. *The lattice of sublattices LS of L*

Of course, there exists a dual structure:

DEFINITION 5.2.– *A subset F of a partially ordered set $(P, \leq)$ is a filter if the following conditions hold:*

1) F is non-empty;

2) F is downward directed: for every $x, y \in F$, there is some $z \in F$ such that $z \leq x$ and $z \leq y$;

3) F is an upper set or upward-closed: for every $x \in F$ and $p \in P$, $x \leq p$ implies that $p \in F$.

From Figure 5.2, we can deduce that the partitions 1, 2, 3, 4, 5 (i.e. $(abc), ..., (ac, b)$) are *ideals* of the lattice LS, while the sequences (1, 4, 5, 3), ... , (1, 3), which are sequences such that $((abc), (ab, c), (ac, b), (a, b, c)), ..., ((abc), (a, b, c))$ are *filters* (properly non-maximal proper filters) of LS. As we have reduced, in this toy-model, philosophies to some partitions, we can identify those philosophies with the *ideals* of the LS-lattice. The other sequences, which are parts of the so-called "philosophies" – or gather structures that belong to more than one – are texts to which, at best, historical and comparative commentaries try to be more or less isomorphic. Therefore, thanks to the LS-structure, and up to an isomorphism, we can easily separate philosophies and commentaries.

We also have another argument to identify commentaries and filters. If each philosophy is a classification whose structure establishes choices or preferences that confer an order on all the elements of the universe on which it is defined, conversely, any commentary relating to the history of philosophy, by the comparisons that he operates between several systems of choices, ultimately comes down to suspending them, at the very moment when he tries to think the network of their coexistences. Any history of philosophy *neutralizes* certain philosophical choices[4]. The one that neutralizes all choices – which is also a theory of the history of philosophy – is identified with the lattice closure of the LS-lattice, in other words, its upper bound.

This *methodological* skepticism, which operates a neutralization of philosophical choices, has nothing to do with the *philosophical* skepticism which is characterized on the contrary by a choice: the most characteristic there is, since it says that "everything is equal" and is formally identified with the discrete partition.

It emerges from this reflection that the "Dianoématique" of M. Gueroult, as a philosophy of philosophy, is not a philosophy strictly speaking, but a variant of history of philosophy, the most imposing without doubt since it is supposed to neutralize "all philosophical choices". This is, of course, a claim since in reality it is a form of idealism – a second-degree idealism.

By this use of lattice formalism, we can thus escape the paradoxes induced by the confusion of levels in Martial Gueroult's "Dianoématique". At the same time, we can see that a philosophy of philosophy is not exactly a philosophy like any other: the LS-lattice includes philosophies (ideals) and, up to isomorphisms, comments (filters) within itself, but does not coincide with any of them. It is a *metastructure*.

4 The (controversial) works of the mathematician Florentin Smarandache, if by chance we wanted to give them credit, are not, for this reason, a "new branch" of philosophy. The famous "neutrosophy" (or study of neutralities), which we will examine in Chapter 10, would be rather a way of formalizing the comments and methodological skepticism specific to historians of philosophy.

Other perspectives might be possible: for example, to re-examine the question of a classification of philosophies, as did Jules Vuillemin in a famous book (see Vuillemin 1984) which attempts to classify philosophical systems according to their conceptions of time. Or still, to consider the systems in a synthetic way, according to some fundamental schemes, and take up again the question of the classification of philosophies (where Kant, Hegel, Cousin, Renouvier and Vuillemin distinguished themselves) from the examination of their classificatory criteria. This supposes that we identify, as Baptiste Mélès does in his thesis (see Mélès 2016), the mathematical form of each classification, which derives, as Gueroult criticized, from the very system of its author. Mélès, however, takes it as a fact that a coherent philosopher can understand of other philosophers only what his own system allows him to think. Thus, the relations between systems are no less liable to structural analysis than the systems themselves. But there is no objectivity in this comparison since each translates the other within its own world. Mélès thus only formalizes Gueroult's intuition, but indefinitely renews the hermeneutics, from which we could nevertheless hope that the appeal to mathematics would bring us out.

In any case, it seems that it would be appropriate to transgress the prohibitions of M. Gueroult and to maintain the possibility of thinking systematically the plurality of philosophies. We give up too quickly, in general, to the facility of the non-system, i.e. the most monstrous system since it does not admit itself as such and is, therefore, unmatchable. It is easy to give in to irrationality or ideology. We can maintain rationality much further with a little tenacity.

5.2. In search of an algebra

Historians of philosophy often explain that doctrines are isolated systems and that we cannot continue another. Thus, Aristotle does not continue Plato, Spinoza or Leibniz do not continue Descartes and Fichte or Hegel do not continue Kant. This is usually true. However, nothing forbids, after all, that different doctrines can be compatible, or that a given doctrine S_3, posterior to two simpler systems S_1 and S_2 cannot, in a way, totalize them. In terms of conceptual architectonics, in any case, the thing is possible. We would like to give an example of this phenomenon by considering the following two graphs K_3 and P_4 (see Figure 5.3).

Let us explain what are P_4 and K_3. The unlabeled path with n vertices is denoted by P_n. A complete graph being a simple graph whose vertices are pairwise adjacent, the unlabeled complete graph with n vertices is denoted K_n. Therefore, P_4 is the unlabeled path with four vertices while K_3 is the complete graph with three vertices.

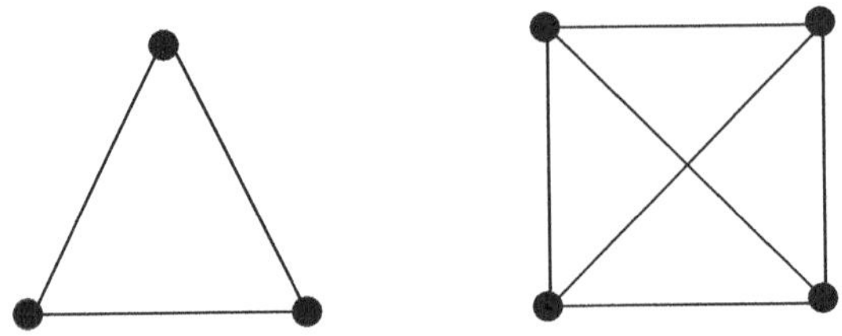

Figure 5.3. *The graphs K_3 and P_4*

Now, we will introduce the following definition:

DEFINITION 5.3.– *The join of simple graphs G and H, written $G \vee H$, is the graph obtained from the disjoint union $G + H$ by adding the edges $\{xy : x \in V(G), y \in V(H)$.*

This can be seen in Figure 5.4.

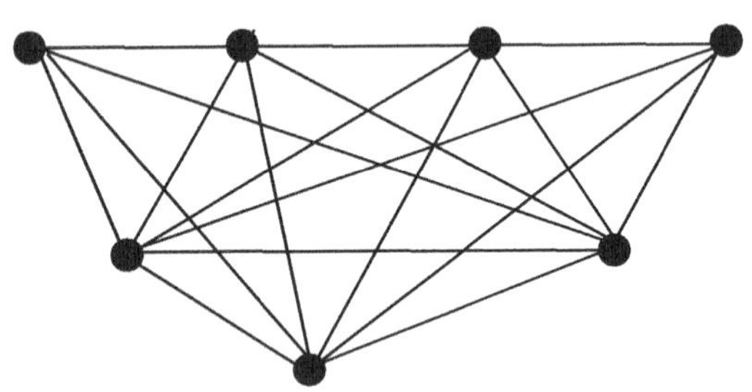

Figure 5.4. *The graph $P_4 \vee K_3$*

In other words, the graph $P_4 \vee K_3$ is obtained by joining each vertex of P_4 to each vertex of K_3.

Assume now that the concepts of a philosophical system are represented by the vertices of some graph, and suppose then two graphs G and H, respectively, associated with the systems S_1 and S_2. When we form $G \vee H$, the join of the two graphs, each concept of S_1 must be connected with each concept of S_2.

Let us now remember that classical metaphysics is traditionally associated with a system of representations equating, as Lachièze-Rey explained, "the self, the world and God", i.e. an instantiation of K_3, the unlabeled complete graph with three vertices. On the contrary, the Heideggerian thought seems to be summed up in the famous diagram of the *Geviert* ("quadriparti", in French), which is the complete set of relations between Earth, Sky, Gods and Men. Therefore, this final graph resembles P_4, the unlabeled path with four vertices. Suppose now we could put together classical metaphysics and Heideggerian thought in one and the same graph: the result would be the join graph $P_4 \vee K_3$ of Figure 6.2.

In linguistics, it has been known for a very long time that certain particular semantic fields (see, for example, the semantic field of seats) can present a mathematical structure (in this case, a Boolean algebra). The Formal Concept Analysis is a sort of complexification of this perspective by exhibiting more general order structures for formal contexts, such as, for example, distributive lattices. The concepts thus take on meaning in relation to each other, according to a partial order fixed once and for all.

6

Concepts and Topology

6.1. Formal concepts

There are some mathematical definitions of concepts.

The most famous, probably, is the definition from Formal Concept Analysis (FCA), a project born around 1980, based on lattice theory, and first carried out by a research group in Darmstadt (see Ganter and Wille 2003). This kind of analysis assumes that the data is structured into units that are formal abstractions of *concepts* of human thought that allow for meaningful and comprehensible information. But the authors immediately warn the reader: "We use the adjective *formal* to emphasize that these formal concepts are mathematical entities and should not be identified with the concepts of human thought" (see Ganter and Wille 2003, p. 591). They also note that the basic data format, the *formal* context, formalizes only an aspect of what is usually referred to as "context".

Using formalism of mathematical orders, data tables can be transformed into algebraic structures called complete lattices, and these can be utilized for data visualization and interpretation. A data table that represents a heterogeneous relation between objects and attributes, tabulating pairs of the form "object o has attribute a", is considered as a basic data type. It is referred to as the *formal context* we speak of above. Now in this theory, a *formal concept* is defined to be a pair (A, B), where A is a set of objects (called the extent) and B is a set of attributes (the intent) such that:

1) the extent A consists of all objects that share the attributes in B, and dually;

2) the intent B consists of all attributes shared by the objects in A.

In this way, Formal Concept Analysis (FCA) formalizes the old semantic notions of *extension* and *intension*. So, we are not so far from what the *logic* of Port-Royal defined as a concept. The main difference is that the numbers of objects and attributes are well defined, as well as the correspondence (Galois correspondence) between the two sets, which are necessarily finite (see Caspard et al. 2007, pp. 92–93; Parrochia and Neuville 2013, pp. 32–36).

In linguistics, it has been known for a very long time that certain particular semantic fields (see, for example, the semantic field of seats) present a mathematical structure (in this case, a binary tree or a Boolean algebra). The Formal Concept Analysis is a sort of complexification of this perspective by exhibiting more general order structures for formal contexts, such as, for example, distributive lattices. The concepts thus take on meaning in relation to each other, according to a partial order fixed once and for all.

Example for a formal context: "bodies of water" (see Figure 6.1):

Attributes

Objects

Bodies of water.	Temporary	Running	Natural	Stagnant	Constant	Maritime
Canal		X			X	
Channel		X			X	
Lagoon			X	X	X	X
Lake			X	X	X	
Maar			X	X	X	
Puddle	X		X	X		
Pond			X	X	X	
Pool			X	X	X	
Reservoir				X	X	
River		X	X		X	
Rivulet		X	X		X	
Runnel		X	X		X	
Sea			X	X.	X.	X
Stream		X	X		X	
Tarn			X.	X.	X	
Torrent		X	X		X	
Trickle		X	X		X	

Figure 6.1. *Example for a formal context: "bodies of water" (source: adapted from Wikipedia article: Formal concept analysis)*

We can, from there, draw up the lattice of concepts in the field of "bodies of water" (see Figure 6.2):

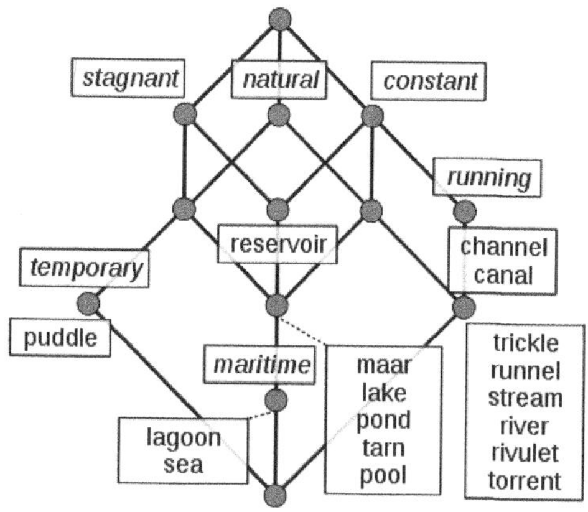

Figure 6.2. *Diagram of concepts corresponding to the formal context "bodies of water" (source: Wikipedia article: Formal analysis concept). For a color version of this figure, see http://www.iste.co.uk/parrochia/graphs.zip*

6.2. Fuzzy concepts

Since Wittgenstein at least, we have known that natural language contains vague elements, which are caught up in a sort of game with respect to each other. Mathematics, here again, has tried to grasp certain aspects of this situation. This is what fuzzy set theory has attempted to do. Knowing that a fuzzy subset is a subset whose membership function of its elements takes its values in [0,1] (for a rigorous formal definition, see Chapter 10, section 10.1), we can introduce, with Kaufmann (1975, pp. 103–104), the following definitions:

DEFINITION 6.1 (FULLY MEANINGFUL WORD).– *Let $x \in T$ and $S(x)$ its meaning or semantic value. If:*

$$\bigvee_y \mu_{\widetilde{S}(x)}(y) = 1, \qquad y \in U, \ x \text{ is said to be fully significant;}$$

$$\bigvee_y \mu_{\widetilde{S}(x)}(y) = 0, \qquad y \in U, \ x \text{ is said to be meaningless.}$$

We can now define a *concept*.

DEFINITION 6.2 ((FUZZY) CONCEPT).– *A fuzzy concept or, more simply, a concept (because, ultimately, a concept is probably always something more or less vague or fuzzy) will, by definition, be a fuzzy subset $\widetilde{A} \subset U$ of a universe of* *discourse U, which, in fact, identifies itself with its meaning or semantic value $\widetilde{S}(x)$. If $x \in T$ is associated with $\widetilde{A}$, we may write:*

$$\widetilde{A} = \widetilde{S}(x) \subset U, \qquad x \in T.$$

To illustrate what we just said, assume that we want to define a concept of morality in a context where we only know a very small population of eight individuals to explain it. We will write, for example:

$$\widetilde{A} = \widetilde{S}(x_1) = \{(y_1|0.3), (y_2|0.9), (y_3|0), (y_4|1), (y_5|0.8), \\ (y_6|0.4), (y_7|0.3), (y_8|0.2)\}.$$

Thus, in the fuzzy set theory, a concept only makes sense if we can specify:

1) the universe of the discourse concerned;

2) the membership function of the concept $\mu_{\widetilde{S}(x)}(y)$ relative to the elements $y \in U$.

It will be noted that this mathematical way of defining a fuzzy concept, i.e. once again, a concept, is obviously much more precise than the intuitive way in which we assign a meaning to the word "concept" in philosophy or in the literary world.

But this precision can also be a disadvantage: on the one hand, in the case of philosophical concepts, it is sometimes difficult to specify the universe of discourse; on the other hand, they do not always correspond to well-determined semantic organizations and cannot enter into finite hierarchies or partial orders.

We will see that philosophers, even when they try to bring some clarifications to their conceptual organizations, remain lax and far too vague most of the time, even when they think they are using a mathematical formalism or pseudo-formalism.

6.3. The case of philosophical concepts

Indeed, philosophers, generally, do not have the mathematics of their concepts: we will illustrate this statement by studying a text that some people would consider, probably, as limpid. It was written by Gilles Deleuze and Félix Guattari in their famous book *Qu'est-ce que la philosophie ?* (What is philosophy?). It begins like this:

> Each concept refers to other concepts, not only in its history, but in its becoming or its present connections. Each concept has components that can in turn be taken as concepts (thus the face has components, but the face itself will be considered as a concept with components). Concepts go to infinity and, being created, are never created from nothing. Secondly, a characteristic of the concept is to render the components inseparable in it: distinct, heterogeneous and yet non-separable, such is the status of the components, or what defines the consistency of the concept, its endo-consistency. It is because each distinct component presents a partial overlap, a zone of neighborhood or a threshold of indiscernibility with another. For example, in the concept of others, the possible world does not exist outside of the face which expresses it, although it differs from it as expressed and expressed; and in turn, the face is the proximity of the words of which it is already the megaphone.
>
> The components remain distinct, but something passes from one to the other, something undecidable between the two: there is a domain ab which belongs to a as well as to b, where a and b 'become' indistinguishable. It is these zones, thresholds or becoming, this inseparability, which define the inner consistency of the concept. But it also has an exo-consistency with other concepts, when their respective creation involves the construction of a bridge on the same plane. Zones and bridges are the joints of the concept.
>
> Thirdly, each concept will therefore be considered as the point of coincidence, condensation or accumulation of its own components (see Deleuze and Guattari 1991, pp. 24–25).

Now we state some proposals for a more formal presentation of these words:

The idea that "each concept refers to another concept" in synchrony and diachrony suggests to a mathematician a graph structure $G(V, E)$, where V is the set of concepts (the vertices of the graph) and E is the set of "references" (the edges or arrows, if these references are oriented, which seems to be the case).

But we later learn that:

1) concepts have "components", which are themselves concepts, etc.;

2) the concepts go to infinity.

We must therefore assume that the set of concepts is an oriented multigraph whose points are themselves graphs, and even graphs of graphs, and this to infinity.

The second point concerns the notion of "consistency of a concept", which apparently presents, for the authors, two aspects, one internal (endo-consistency) and the other external (exo-consistency).

Endo-consistency seems linked to the fundamental notion of "a component of a concept", which immediately raises the question of the underlying topology of the set of concepts, with its neighborhood relations.

To give meaning to the text, we cannot avoid making $G(V, E)$ a topological space, i.e. providing V (the set of vertices) or E (the set of edges) of a topological structure[1].

Recall the following definition:

DEFINITION 6.3 (TOPOLOGICAL SPACE).– *According to Bourbaki, a topological structure (or more briefly a topology) on a set X, is a structure constituted by the data of a set O of subsets of X having the following properties:*

(O1) any union of sets of O is a set of O;

(O2) any finite intersection of sets of O is a set of O.

The sets of O are called "open sets" of the topological structure defined by O on X, and a topological space is called a set with such a topological structure.

In the case of the concepts, this topology, however, does not seem to be that of a so-called "separable" space. Indeed, we are told that "the characteristic of a concept is to make the components inseparable in it" and that "each distinct component has a partial overlap, a neighborhood zone or a threshold of indistinguishability with another". In the concept of *Others*, we explain, for example, the face, the possible world that it expresses, and the words of which it is the mouthpiece are non-separable components.

How do we take these requirements into account?

1 Obviously, it is not the same thing if it is V or E. We will not have the same topological space at all.

We know that in topology we distinguish so-called separated spaces (or Hausdorff spaces) and non-separated spaces. What is a separated space? To understand this, first define the concept of a neighborhood.

DEFINITION 6.4 (NEIGHBORHOOD).– *In a topological space X, we call the neighborhood of a subset A of X any set that contains an open set containing A (the neighborhoods of a subset $\{x\}$ reduced to a single point are also called neighborhoods of the point x).*

Now, we know that there are several equivalent properties that characterize separated spaces. Let us take the simplest and most intuitive of them:

DEFINITION 6.5 (THE HAUSDORFF SPACE).– *A topological space X is "Hausdorff" if, whatever two distinct points x and y of X, there is a neighborhood of x and a neighborhood of y without a common point.*

Example: a discrete space is separated. Take, for example, the rational line $\mathbb{Q}$. If x and y are two rational numbers such that $x < z < y$, the respective neighborhoods $]\leftarrow, z[$ and $]z, \rightarrow[$ of x and y do not meet.

A topological space X is said to be non-separated if, whatever the distinct points x and y of X, there is not a neighborhood of x and a neighborhood of y without a common point.

Example: let T be the least fine topology on a set X, namely the topology whose only open sets are $\emptyset$ and X. If X has two points a and b, then a neighborhood of a and a neighborhood of b necessarily have to intersects the set X, since:

$$V(a) = \{\{a\}, \{b\}\} = \{X\} = V(b).$$

Similar examples can be found in quantum mechanics, and also in some existential situations or quite ordinary experiments[2]. Consider, for example, the set of colors. We know that from three basic colors, we can form, by composition, other colors. There is therefore no uniqueness of the basic colors, as long as a subfamily of orthogonal colors generates all of the others. Thus, if *blue, yellow* and *red* form a base of colors, *green* does not separate from *blue* and *yellow* (see Sallantin 1986).

Let us now return to the topology of the concept components of the DG space (DG stands for Deleuze–Guattari). Knowing that every component of a concept is

2 We have given examples of this in a previous book (see Parrochia 1991).

a concept, we can easily show that, in such a space, for any pair of components, we can always find a third component, such that the first one is in the concept created from the second and the third.

Does this mean that, in such a space, every neighborhood is necessarily a non-separable one?[3] No. It will be noted that if the non-separability of neighborhoods necessarily induces the non-separability of space, conversely, the non-separability of the topological space does not necessarily induce a non-separability of neighborhoods. Non-separated spaces are known in which every point has a separated neighborhood.

Example: let X be an infinite set. The set of subsets of X formed by X and $M \subset X$ such that Card(M) $< c$ is the set of closed subsets of X for a T_0 topology. We know then that if $c >$ Card(X), T_0 is the discrete topology, which is obviously separated. There are, of course, separated topologies on X for which there are countable infinite sets that are not closed. Hence, the intersection of all the separated topologies on X is the non-separated topology $T_{Card(\mathbb{N})}$ which is also the least accessible topology on X.

The topology of the DG space is therefore a non-separated space topology. What are the consequences? To find them out, we need to introduce another important notion in topology: compactness.

DEFINITION 6.6 (COMPACTNESS).– *We say that a topological space X is quasi-compact if it satisfies one of the following axioms:*

1) every filter on X has at least one adherent point;

2) any ultrafilter on X is convergent;

3) any family of closed sets in X whose intersection is empty contains a finite subfamily whose intersection is empty;

4) any open covering of X contains a finite open covering of X.

A space that is quasi-compact and separated is a compact space.

We know that every finite, separated space is discrete and that every finite, discrete space is compact. Conversely, we can show that every discrete, compact space is finite.

---

3 Deleuze and Guattari, who do not know mathematics, seem to presuppose it when they write: "we started from a rather complex example: how can we do otherwise, since there is no simple concept?"

The DG space is unfortunately not a separated space, and it is an infinite space. It can therefore be, at most, only quasi-compact.

Indeed, let X be the set of concepts. Let us equip X with the topology for which the closed subsets are X and the finite subsets of X. X satisfies the axiom 3 and, since it is infinite, it is not separated.

Note that in a non-separated space X, a compact set A is not necessarily closed and its *adherence* (i.e. the set of points x whose neighborhood $V(x)$ meets A) is not necessarily quasi-compact.

Likewise, the intersection of two compact sets is not necessarily quasi-compact, and the union of two compact sets is not necessarily compact.

As a result, conceptual decompositions and recompositions are not necessarily achievable in DG. Point 3 will therefore require, to be satisfied, an additional structure on DG.

DEFINITION 6.7 (CONNECTEDNESS).– *A topological space X is said to be connected if there is no union of two disjoint non-empty open sets (see Bourbaki 1971, 80, §11). The connected component of a point in a space X is the largest connected subset of X containing that point. The connected components of the points of A, referred to as the subspace A of X, are called connected components of a subset A of X.*

DEFINITION 6.8 (DISCONNECTED SPACE).– *We say that a space X is totally disconnected when the connected component of each of its points is the set reduced to this point. A discrete space is naturally totally disconnected but the reciprocal is not exact.*

These clarifications being made, what can be said about DG's connectedness and that of its conceptual "components"? The text assures us – it is apparently this truism that defines "exo-consistency" – that concepts are interrelated if their respective creation implies the existence of a link in a certain plane.

It should first be remembered that concepts are graphs of graphs whose component topology is that of a non-separable space.

We will admit that the components of the concepts are related, although this property does not derive from non-separability. On the contrary, is the concept space itself connected? If the concepts are the open subsets of the space, two concepts, that is, two open non-empty subsets, can be not disjoint and, if this is the case, their link belongs to a certain hyperplane. Under what conditions could the X-space of concepts be said to be connected? It would be so if (at choice):

– there exists, apart from X and the set $\emptyset$, no subset of X that is both open and closed;

– there is a set A included in X, connected and everywhere dense.

Indeed, reason by the absurd. If X is not connected, there exist two non-empty disjoint open sets N and M, such that $X = M \cup N$. But then, $M \subset A$ and $B \subset A$ are, in A, two non-empty open sets with no common points, whose union is A, which makes A a non-connected subset, contrary to the hypothesis.

The question is: knowing that a concept is a connected part A of X, can we know if it is everywhere dense?

Recall that a part A of a topological set X is everywhere dense if its adherence is none other than the set X, in other words, if, for any non-empty open subset U of X, $U \cap A$ is non-empty.

The question then becomes: does every non-empty open subset of x, i.e. every concept, intersect a concept? In other words, does X contain regions nowhere dense, concepts or sets of concepts totally isolated from others? The text is silent on the question, perhaps because it is a place of confrontation between theories: for example, the space of the concepts is certainly totally connected for Heidegger or Derrida, while is surely not connected for Bachelard, Canguilhem or Foucault.

For want of clarifying this point, or of explaining why the authors refuse to specify it, the text remains philosophically totally ambiguous. Note that the existence of a non-connectivity on the conceptual space deprives beautiful results.

In particular, we know that every quotient space of a connected space is connected. Assuming that the conceptual space is connected, we can deduce the connectivity of the conceptual equivalence classes. In other words, the connectivity of the philosophical conceptual space could make it possible to prove the connectivity of the philosophies understood as subspaces. On the contrary, assuming that the space X is not only connected but also locally connected, then, some beautiful theorems of topology would apply to this space, like the Poincaré–Volterra theorem, which makes it possible to decompose X into a countable family of open sets, each of which is contained in a part of a set B, a set of parts of X whose interiors form an overlap.

In other words, the insertion of each concept into other broader concepts covering space could result from the connectivity and local connectivity of conceptual space. From the hypotheses posed, we could thus almost base the legitimacy of descriptions or comments conceived as sets of concepts containing the concepts existing in X (e.g. the history of philosophy for philosophy). But here: if the conceptual space DG is – which seems probable – not connected, it follows that, if $\mathcal{R}$ is the relation "y belongs to the connected component of x", R is an equivalence relation and the X/R quotient space is totally disconnected!

We must now address the third point: the idea that a concept can be considered as the "point of coincidence, accumulation or condensation of its components".

This vocabulary is obviously inappropriate. In particular, the notion of "accumulation point" is used in mathematics as follows. Let us give $\mathbb{R}$, the set of reals and a part A of $\mathbb{R}$. We say that b belonging to $\mathbb{R}$ is an accumulation point of A if every open interval V containing b contains infinitely many points of A. We can show that every bounded infinite set of real numbers has at least one accumulation point. This is another version of Bolzano–Weierstrass's theorem which says that for any bounded infinite sequence of real numbers, we can extract a convergent sequence[4].

Let us look for the equivalent of such a proposition in the case raised by the text. Let us give a topological space X and $(x_n), n \in \mathbb{N}$ an infinite sequence of points that converges to a point b. We can easily prove that the set a formed of $x_n \in \mathbb{N}$ and b is quasi-compact.

The authors would therefore in fact want the space DG, not separated and not compact, to contain only quasi-compact subsets A, i.e. such that any overlap of A by open sets in X contains an overlap finished by A. But in addition, they would like this DG space to be equipped with convergent suites.

4 As shown by Fréchet, who introduced the notion of compact space, this theorem is very close to the Borel–Lebesgue theorem. Riesz is also to be cited, in this regard, among the precursors of topology, since he started precisely from the notion of point of accumulation or, what comes to the same thing, of a derived set. But it was Hausdorff, Alexandroff, Urysohn and Tychonov who developed the theory of compact spaces, before Cartan – inventor of the notion of "filter" and the very general definition of convergence, thanks to the theorem on ultrafilters – came to simplify the theory. The informal, and moreover probably whimsical, use of the notion of "accumulation point" by Deleuze–Guattari refers, at best, to a completely bygone pre-topological era. Indeed, if (U_i) is a covering of A by open sets of X, there exists an index k such that b belongs to U_k. As U_k is a neighborhood of b, there only exists a finite number of indices n_k, such that x_{n_k} does not belong to U_k. If, for each index k, t_k is an index such that $x_{n_k w} \subset U_{t_k}, U_k$ and the t_ks form a finite open covering of A (see Bourbaki 1971).

This amounts, in fact, to establishing a uniform structure on X, then transform the space X into complete space, i.e. to establish a filter base of Cauchy on it (see Bourbaki 1971, pp. 1–12).

A special case is where X is a discrete uniform space, since a Cauchy filter on X is then a trivial and therefore convergent ultrafilter. As a result, X is complete, and any sequence of Cauchy is necessarily convergent (see Bourbaki 1971, p. 15).

The X-space of the concepts, conceived independently of their U-reference, as a discrete uniform space provided with Cauchy filters, would thus already answer the question. In such a case, it would follow that we would not want, contrary to what happened in Plato, to "compact" (see Parrochia 1993a) space by any point at infinity (Good, God, any transcendent element) or project into any compact space (Genres). The reasoning can take place, as the authors undoubtedly wished it, entirely in the "plane of immanence". However, we see what formal conditions of this philosophical requirement can be fulfilled. Its satisfaction requires, given the presuppositions of departure, the establishment of a non-trivial structure over all the concepts, with, in particular, a condition of convergence, and we would like to know what justifies it. Since the authors are dead and probably far from imagining what their theory of philosophy presupposed, the question will have no answer.

On the contrary, this brief reflection on an example of a mathematical structure likely to be associated with a philosophical conceptual space would, in line with a previous book on classical philosophical structures (see Parrochia 1993a), draw the attention of our colleagues to formal conditions of the exercise of our activity and perhaps to sketch the idea of a systematic classification of philosophical conceptual spaces as to their separability, compactness, connectivity, finiteness, continuity, etc.[5].

All combinations are not possible: thus, if a space is discrete, infinite, it is necessarily not compact.

Where philosophers usually do not see any difference or constraint on thought, mathematics can bring out very real distinctions, where much of what we can think is already played out.

We can of course, as analytical philosophy has done, track down the philosophical error in the argumentation: it is often a bad trial because the

5 It goes without saying – but that is another story – that the origin of these notions deserves to be traced and epistemologically analyzed.

strength of a philosophy does not reside, generally, in the proof that it brings what it advances, but in the fantastic conceptual creation that characterizes its universe and makes it totally distinct from another. This leads to concern also, and perhaps most importantly, for the conceptual architectures that philosophers produce. One of the great advances of structuralism, which resulted from a rigorous mathematical work before becoming the approximate toy of the human sciences, consisted of highlighting very general formal constraints, evidently at work in languages, societies, minds and behaviors. It would be decidedly light that the wear of words and the reluctance of the air of time to any form of rationality, even elementary, lead to throwing the baby out with the bathwater.

7

The Problem of the Infinite

What about infinity? As we have seen, a philosophical system is necessarily expressed in a finite language. Philosophy is very redundant: constructing a philosophical system needs only a few number of categories. But the final aims of a philosophy may be sometimes the same as those of religion: to reach the infinite. Even in pantheistic philosophies, such as those of Spinoza or Whitehead, where the world is more or less identified with God, there is still some trace of the infinite in this world.

For a long time, reflection on infinity remained the prerogative of philosophers and theologians. But this definitely ended in the 19th century.

The first mathematician to be really concerned with the problem of the infinite is Cantor.

Working on sets, Cantor, as we know, is at the origin of the distinction between two types of infinites: the countable, represented by the cardinal number $\aleph_0$, and the continuum c, which cannot be bijected with the set of natural numbers (the famous diagonal process shows that there will always be at least one excess number). The question then arises of understanding what the continuum corresponds to. For Cantor, the continuum was none other than $\aleph_1$, and the passage between successive alephs, generalization of the passage from the cardinal of a set to the cardinal of the set of its subsets, was defined by:

$$\aleph_1 = 2^{\aleph_0} \quad \text{(continuum hypothesis or CH);}$$

and, for every $\alpha \in \mathbb{N}$:

$$\aleph_{\alpha+1} = 2^{\aleph_\alpha} \quad \text{(generalized continuum hypothesis or GCH).}$$

Cantor tried unsuccessfully to demonstrate these assumptions throughout his life. We know today, from the independence results of Gödel and Cohen, that it is not possible to prove CH or its negation inside ZFC.

So, after Cantor, and especially after Gödel and Cohen, the infinite has lost its simplicity. Not only can the infinite be discrete or continuous but also the continuum itself can be identified with any aleph, provided that it is not the aleph corresponding to the countable.

The independence results and the definitive insolvability of certain mathematical problems within the framework of ZFC led to two conflicting positions: the first, of which can be called "pluralism", maintains that this situation makes the enterprise of building a set theory largely indeterminate. As there can now be several set theories (maybe an infinite number), the choice between one or the other can, of course, be for practical reasons, but not for theoretical ones. The second position, which we will qualify as "non-pluralism" rather considers that the situation testifies above all to a lack (the poverty of the resources available to us to justify mathematical statements). Consequently, it is advisable to seek reasons capable of justifying the introduction of new axioms, which amounts to saying that there is an objective mathematical universe and only one, which would be truly of importance for philosophy.

To go further, it is first necessary to recall some elements of the arithmetic of infinite cardinals.

7.1. The arithmetic of infinite cardinals

The addition and multiplication of infinite cardinal numbers is trivial: for infinite cardinals κ and λ:

$$\kappa + \lambda = \kappa.\lambda = \max(\kappa, \lambda).$$

The situation becomes interesting when we turn to exponentiation and the attempt to compute $\kappa\lambda$ for infinite cardinals.

During the dawn of set theory, Cantor showed that for every cardinal κ:

$$2^{\kappa} > \kappa.$$

This strict inequality means that a set E cannot be bijected with $P(E)$, the set of its subsets, because $|P(E)| = 2^{E}$, the cardinal of $P(E)$, is strictly greater than $|E|$, the cardinal of E.

This explains the cantorian hierarchy of sets V and also his famous conjectures: the continuum hypothesis CH and the generalized continuum hypothesis (GCH).

But as the latter could not be proven, other concepts and structures have been introduced to go on in the exploration of the infinite. Maybe one of the most important is the notion of cofinality.

Consider a set A endowed with a binary relation $\leq$. A subset B of A is said to be *cofinal* if, for any element a of A, there exists an element b of B such that $a \leq b$. In symbols:

$$\forall a \in A, \exists b \in B \mid a \leq b.$$

The cofinality of set A is the cardinality of the smallest cofinal subset of A.

The cofinality of a limit ordinal α is the smallest ordinal β such that there is an unbounded function $f : \beta \to \alpha$. This ordinal is usually denoted $\mathrm{cf}(\alpha)$.

Intuitively, $\mathrm{cf}(\alpha)$ is the fewest steps to take to get to the end of α.

For example, we can go to the end of $\aleph_0$ in $\aleph_0$ steps, with the identity function, but we cannot go to the end of $\aleph_0$ in a finite number of steps. So we have $\mathrm{cf}(\aleph_0) = \aleph_0$.

A cardinal that is equal to its cofinality, like here, $\aleph_0$, is called a *regular* cardinal.

Likewise, we can go after $\aleph_1$ in $\aleph_1$ steps, but we cannot do it in a countable number of steps. We therefore have $\mathrm{cf}(\aleph_1) = \aleph_1$ and $\aleph_1$ is therefore also a regular cardinal.

On the other hand, we can go to the end of $\aleph_\omega$ in $\aleph_0$ step, with the function $f : \aleph_0 \to \aleph_\omega$ defined by $f(n) = \aleph_n$, so $\mathrm{cf}(\aleph_\omega) = \aleph_0$.

A cardinal which is not regular, i.e. which is not equal to its cofinality, as here $\aleph_\omega$, is called a *singular* cardinal. For example, $\omega + 1$, which is the next ordinal number greater than ω, is singular since it is not a limit ordinal. Let us examine $\omega + \omega$. It is the next limit ordinal after ω, and it can be written as the limit of the sequence:

$$\omega, \omega + 1, \omega + 2, \omega + 3..., \text{etc.}$$

This sequence has an order type ω, so $\omega + \omega$ is the limit of a sequence of type less than $\omega + \omega$, whose elements are ordinals less than $\omega + \omega$; therefore, it is singular[1].

As far as we know, only the aleph numbers can meaningfully be called regular or singular cardinals[2]. But there exist some others.

7.2. The question of large cardinals

In the total universe of infinite sets, we do not only encounter regular or singular cardinals. There are also sets which are called the "large cardinals". What is a large cardinal?

The answer is not obvious. In fact, according to the authors, we can find different definitions. To fix the ideas, let us say first that a large cardinal is a transfinite cardinal number satisfying a property that distinguishes it from sets constructible with usual axiomatics (ZFC) such as $\aleph_0, \aleph_\omega$, etc., and makes it necessarily larger than all of them. The existence of a large cardinal is therefore subject to the acceptance of new axioms.

So the question shifts to this one: since large cardinals are defined by axioms, what is a "large cardinal axiom"? Of course, we can roughly say that a large cardinal axiom is an axiom stating that there exists a cardinal (or perhaps many of them) with some specified large cardinal property. But what are the properties in question?

Again, the answer varies from book to book. As a first approximation, we can draw up the following table, which arranges cardinals in order of consistency strength, with size of the cardinal used as a tiebreaker. But in some cases (such as strongly compact cardinals), the exact consistency strength is not known and the table uses the current best guess. So, beyond the "small" cardinals

1 For a complete presentation of transfinite arithmetic, see Jech (2003), and for an excellent summary, see Petitot (2009), pp. 5–7.

2 However, we must note that uncountable (weak) limit cardinals that are also regular are known as (weakly) inaccessible cardinals. They cannot be proven to exist within ZFC, though their existence is not known to be inconsistent with it. So their existence is sometimes taken as an additional axiom. Inaccessible cardinals are in fact fixed points of the aleph function, though not all fixed points are regular. For instance, the first fixed point is the limit of the ω-sequence $\aleph_0, \aleph_{\aleph_0}, \aleph_{\aleph_{\aleph_0}}, ...$ and is therefore singular.

$0, 1, 2, ..., \aleph_0, \aleph_1, ..., \kappa = \aleph_\kappa$, beyond the so-called "wordly cardinals"[3], we can find the following ones[4]:

1) weakly and strongly inaccessible, α-inaccessible and hyper-inaccessible cardinals;

2) weakly and strongly Mahlo, α-Mahlo and hyper-Mahlo cardinals;

3) reflecting cardinals;

4) weakly compact (= Π_1^1-indescribable), Π_n^m-indescribable, totally indescribable cardinals;

5) λ-unfoldable, unfoldable cardinals, v-indescribable cardinals and λ-shrewd, shrewd cardinals (not clear how these relate to each other);

6) ethereal cardinals, subtle cardinals;

7) almost ineffable, ineffable, n-ineffable, totally ineffable cardinals;

8) remarkable cardinals;

9) α-Erdös cardinals (for countable α), $0^\#$ (not a cardinal), γ-iterable, γ-Erdös cardinals (for uncountable γ);

10) almost Ramsey, Jónsson, Rowbottom, Ramsey, ineffably Ramsey, completely Ramsey, strongly Ramsey, super Ramsey cardinals;

11) measurable cardinals, $0^\dagger$;

12) λ-strong, strong cardinals, tall cardinals;

13) Woodin, weakly hyper-Woodin, Shelah, hyper-Woodin cardinals;

14) superstrong cardinals (=1-superstrong);

15) subcompact, strongly compact (Woodin $<$ strongly compact, $\leq$ supercompact), supercompact, hypercompact cardinals;

3 In set theory, a worldly cardinal is a cardinal κ such that the rank V_κ is a model of Zermelo–Fraenkel set theory (Hamkins 2014). By Zermelo's theorem on inaccessible cardinals, every inaccessible cardinal is worldly. But by Shepherdson's theorem, inaccessibility is equivalent to the stronger statement that $(V_\kappa, V_{\kappa+1})$ is a model of second-order Zermelo–Fraenkel set theory (see Kanamori 2005, p. 19). So, being worldly and being inaccessible are not equivalent; in fact, the smallest worldly cardinal has countable cofinality and therefore is a singular cardinal (see Kanamori 2005, p. 57).

4 We just enumerate here some known large cardinals and do not want to define them in details (it would be too technical). It is sufficient to know that there are, roughly speaking, three categories classified according to the power: measurable cardinals, strong and superstrong cardinals and, in between, a category introduced by Woodin and named this "Woodin cardinals".

16) η-extendible, extendible cardinals;

17) Vopěnka cardinals, Shelah for supercompactness, high-jump cardinals;

18) n-superstrong ($n \geq 2$), n-almost huge, n-super almost huge, n-huge, n-superhuge cardinals (1-huge = huge, etc.);

19) wholeness axiom, rank-into-rank (axioms I3, I2, I1 and I0).

We can add to this list even stronger large cardinals that are not consistent with the axiom of choice, but whose existence has not yet been refuted in ZF alone (i.e. without use of the axiom of choice). These are the Reinhardt cardinal and the Berkeley cardinal.

With this enumeration being made – there would in fact be several dozen axioms of major cardinals – we can try to propose a generic definition. In view of the above, we could suggest that a large cardinal be considered such if it satisfies some open-ended, semi-formal "definition", founded on the fact that, historically, large cardinal assumptions seem to fall into one of several categories:

1) inaccessibility "from below" of some sort;

2) critical points of elementary embeddings between certain set-theoretic structures;

3) propositions formalizable in, say, ZFC or some second-order strengthening/alternative, which logically entail the existence of other large cardinal notions.

Per 1: This criterion will obviously capture the weakly and strongly inaccessible large cardinals as closure points under the usual set-theoretic operations. Maybe this informal notion will also give us Mahlo cardinals. We would also include in this category indescribable cardinals of various degrees, hence also weakly compact cardinals (though these will fall under both categories 2 and 3 as well). This inaccessibility requirement also encompasses proof-theoretic considerations. So, for example, statements that entail the existence of models of say ZFC would fall under this category as well. This would include the iterated consistency hierarchy, the existence of a transitive model of ZFC and worldly cardinals and their generalizations. The question may arise as to whether to put in this category the theory $V = L$ or, at the very least, the desired variant of it which would be supposed to be compatible with large cardinals.

Per 2: This particular category encompasses, as far as we know, most of the large cardinal axioms and, since it is pretty standard, should not, in principle, pose too many problems.

Per 3: This last category is intended to encompass assumptions relating to the existence of various kinds of sharps, indiscernibles for various set-theoretic structures, and axioms like $AD^{L(\mathbb{R})}$[5] and generalizations of these structures. This category could also encompass the existence of certain types of indestructible large cardinals since many such cardinals have been shown to exist by assuming the existence of a particularly large cardinal and then providing a certain forcing construction. Of course, this criterion may be a little unfair since the proposition 0=1, as seen in Kanamori (2005, p. 472) implies every set-theoretic statement, including all so-called large cardinals.

Of course, there are numerous other attempts of definitions for large cardinals.

Consider, for example, the definition given by Woodin.

DEFINITION 7.1.– *$\exists x\, \phi(x)$ is a large cardinal axiom, if $\phi(x)$ is a Σ_2-formula, and as a theorem of ZFC, if κ is a cardinal such that $V \models \phi(\kappa)$, then κ is strongly inaccessible, and for all forcing notions P of size $s < \kappa$, $V^p \models \phi(\kappa)$.*

This definition raises a lot of questions because it does not cover all possible situations: for example, it seems that it does not cover cardinals like supercompact, strong, strongly compact, etc. since these are not Σ_2-definable. But we can understand this point of view. Actually, Woodin's definition is in the context of a proper class of Woodin cardinals, so it is definitely not meant to be comprehensive (or to be taken seriously as an attempt to formalize the notion). Rather, it is an attempt to formalize our intuition that the large cardinal hierarchy is indeed well-ordered, by showing that a reasonable subclass is well-ordered under appropriate assumptions. The restriction to Σ_2 is to make the notions local. Supercompactness, or even strongness, is not of this kind.

First attempts to define rigorously large cardinals amount probably to Penelope Maddy's papers (1988a, 1988b), developed in some of her books (1997, 2011) generally well received by the scientific community and considered fairly faithful to the experience of the "working mathematician". A general axiomatic framework for large cardinal axioms has been also attempted by Apter et al. (1989) at the end of the 1980s. We probably could say more simply that a large cardinal axiom is an axiom assuming the existence of a κ so that $\kappa \geq \text{Min}(\lambda : \lambda = \aleph_\lambda)$. As such, we would have to consider the existence of aleph-fixed point to be the "smallest" large cardinal axiom.

5 AD is the axiom of determinacy and $L(\mathbb{R})$ is a particular structure based on Gödels hierarchy L, which the mathematician Hugh Woodin hopes will contribute to prove the continuum hypothesis. See section 7.3.

It is likely that none of these definitions is general enough to encompass all existing or future great cardinal axioms. But, if nothing else, they still give an idea of what these objects are.

Now, if we try to classify these different properties, we encounter, according to the authors, different orderings. Let us give a quick sample below: the charts of cardinals from A. Kanamori (2005, p. 472) range from 0–1 down to weakly inaccessible cardinals in this way:

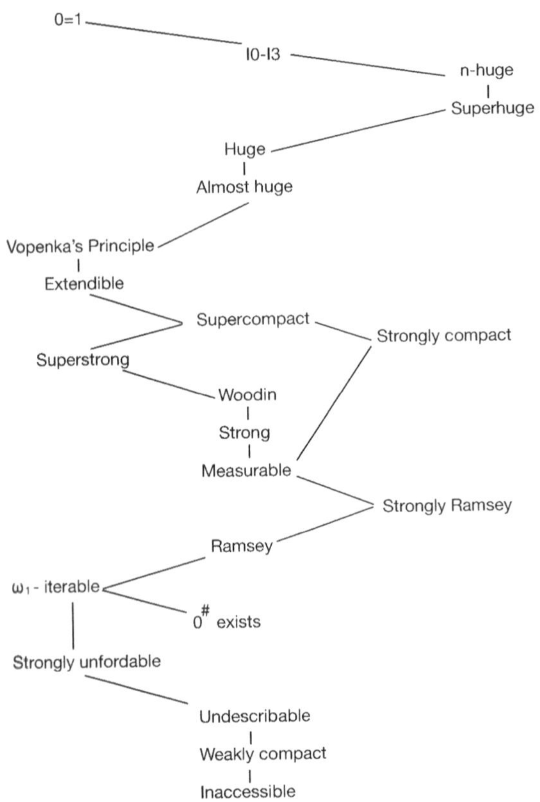

Figure 7.1. *Chart of large cardinals (after Kanamori 2005, p. 472)*

But there are other types of ordering in the literature, generally linked to specific contexts: for example, that of Ramsay's cardinals for Victoria Gittman (2007) or that of more specific regions of major cardinals for Norman Perlmutter (2013). Some of these diagrams overlap, such as that of Dan Saatrup Nielsen (see Nielsen 2018; Dimopoulos et al. 2021) and that of Kanamori, others do not.

The problem now is to see how large cardinals intervene in attempts to prove the continuum hypothesis. It is therefore now a question of examining the most important advance of recent years on the question, namely Woodin's program, the philosophical consequences of which are major.

7.3. Woodin's program

As we have seen, Cohen's work did not necessarily end the debate about the continuum hypothesis: there remains the possibility of discovering new "plausible" axioms that resolve the question one way or the other. Certainly, Cohen himself showed that the axioms of large cardinals alone cannot modify the undecidability of the continuum hypothesis, but in works from the early 2000s (see Woodin 2001a, 2001b), W. Hugh Woodin envisaged that the continuum hypothesis can be essentially false by introducing a meta-logic called Ω-logic[6], based on projective sets.

Woodin's Ω-conjecture said that any essentially true statement in Ω-logic is Ω-provable. Using axioms of large cardinals, plus the above conjecture, he then deduced that the continuum hypothesis would be essentially false, and more precisely that $2^{\aleph_0}$ would be equal to $\aleph_2$ (hypothesis already considered by Gödel).

At that time, these results were far from unanimous among theorists (see Dehornoy (2003) and also Bellotti (2005)). But, from 2010, a spectacular reversal has occurred. Woodin, exploring a variant (compatible with the axioms of large cardinals) of the axiom of constructibility, identifies the class of sets V with an extended constructible universe, "Ultimate L", leading on the contrary to models where the continuum hypothesis is "naturally" true.

It may seem extremely disappointing that, within a few years, a great mathematician can deny himself. Without going into too much detail, let us just try to understand the approach that led Woodin to successively adopt one thesis, then the other.

7.3.1. *Woodin I: CH would be false*

We will be largely inspired here by Dehornoy (2007). Let us start with the notion of *forcing*. To show that the ZFC system did not prove the continuum hypothesis (CH), Cohen left the initial model of the theory to build an extension

6 This notion has been well explained by Patrick Dehornoy (see Dehornoy 2007).

that did not satisfy CH. Unfortunately, this very powerful method of forcing – which notably made it possible to show that many other questions, such as the axiom of choice or Suslin's hypothesis, are not decidable in ZFC – has its downside: thanks to this, we can just as well define a forcing resulting in a model extension satisfying CH. We thus obtain a complete symmetry – or indiscernibility – between the continuum hypothesis and its negation.

It is this situation that Woodin wanted to change. Banking on the fairly certain character of arithmetic (Gödel's incompleteness theorems not being a really limiting factor for the possibilities of demonstration), he sought to find the situation of ZFC and the arithmetic, satisfactory in practice, for larger fragments of the universe of sets. In this context, an axiomatic system will be a *good solution* for a fragment H of the universe of sets if this system makes the properties of H invariant by forcing, i.e. if it neutralizes the action of forcing on H. In other words, if H has a certain property P in a model of ZFC, it necessarily keeps this property in any extension by forcing the model considered. Therefore, if any solution for H entails such a property P, we will say that P is established, or alternatively, that it is *essentially true*.

Let us now call H_{Ω_0} the domain of the finite world, H_{Ω_1} that of countable infinity, H_{Ω_2} that of cardinality at most $\aleph_1$, etc. Arithmetic should be considered the correct solution for H_{Ω_0}. The search for solutions for the not at all obvious fragment H_{Ω_1} was the main task of set theory between 1970 and 1985. The culmination of this search was the existence of a good solution for H_{Ω_1} in the form of the system obtained by adding to ZFC the axiom called "projective determination". This solution was established by Woodin in 1984, and the Martin–Steel theorem later showed that the axiom of projective determination was in fact itself a large cardinal axiom, since it is essentially equivalent to the existence of an infinity of cardinals called "cardinals of Woodin". For H_{Ω_2}, the problem is more complicated because the large cardinal axioms cannot constitute solutions and other ones must be found. Some candidates have been spotted, which are extensions of Baire's theorem stating that a countable intersection of dense open sets of the real line is never empty. Such are, for example, Martin's axiom or, in its strong form, Martin's axiom maximum, which Shelah has shown gives a fairly good description of the fragment H_{Ω_2}. Foreman, Magidor and Shelah had also proven in 1988 that it was compatible with the existence of the great cardinals.

A variant is Woodin's Maximum Martin axiom, denoted here MMW. This axiom guarantees the invariance of the properties of $H\Omega_2$ by forcing and provides a very complete description of this fragment. On the other hand, the compatibility between MMW and the existence of large cardinals remained an open problem, so that it was impossible to assert the ZF+ MMW system as a solution for H_{Ω_2}.

It was then the fact that Woodin had the idea of introducing a new conceptual framework, in this case a new logic baptized "Ω-Logic", directly integrating in its construction invariance by forcing and being compatible with the existence of great cardinals. In Ω-Logic, proofs are particular subsets of the real line, called "universally Baire"[7], and not, as in usual logic, sequences of statements obeying syntactic rules. As in usual logic, a proof in Ω−Logic is a certificate guaranteeing that a statement has a certain property. Here, a universally Baire set A is said to be an Ω-proof for a formula ϕ if ϕ is true in all countable ZFC models such that A remains universally Baire in any forcing extension of the model. Any formula provable in usual logic is also provable in Ω-Logic, but the converse is false because there exist universally Baire sets which are much more complicated than ordinary Borelians, and Ω-provability does not guarantee provability in the usual sense. As for validity in Ω-Logic, it is defined with reference to the canonical model formed by true sets endowed with the true membership relation: a statement is said to be Ω-valid if it is true in any extension by forcing of this model.

Woodin shows that Ω-Logic is consistent, in the sense that any Ω-provable statement is Ω-valid. Several arguments suggest that the reciprocal implication is also verified, i.e. that the Ω-Logic is complete, but this remains a conjecture. We then come to the goal of all this construction.

In 2000, Hugh Woodin proved that any Ω-complete axiom for H_{Ω_2} implies the fallacy of the continuum hypothesis, an extraordinary tour de force since it does not only concern axioms such as MM or MMW but any axiom of the same species already introduced or to come. We can then express the conjecture as follows:

CONJECTURE 7.1 (WOODIN 2000).– *Unless the Ω-conjecture is false, the continuum hypothesis is essentially false.*

7.3.2. *Woodin II: CH may be true*

We were there in the early 2000s. But the situation has turned around. Why? To understand this, the best thing to do is to follow here the beautiful article by Colin J. Rittberg (2015). The question was to prove the Ω-conjecture and Woodin seemed ready, until then, to analyze the whole universe fragment by fragment.

7 This notion was introduced in Feng et al. (1992).

But in 2010, the perspective changes. It appeared that the Ω-conjecture might be refutable. After all, it was enough to find a forcing extension of V where the conjecture does not hold to refute it. It was therefore not unreasonable to expect both that the Ω-conjecture had an answer and further, that this answer was that it was false, in particular that it was refuted from a large cardinal hypothesis.

Obviously, the level of such a cardinal in the hierarchy of cardinals should be rather large and notably beyond the level represented by a certain type of internal model. Conversely, if the Ω-conjecture were true, then there would be no such critical level. So, answering this question – is there a critical level or not? – was then Woodin's new motivation to start a project in internal model theory, the results of which can be found largely in Woodin articles of 2010 and 2011. It turned out that there was evidence (but no hard evidence) to support that there was no such critical level, and that in fact the Ω-guess may well be true.

However, in the process of gathering evidence for this, something else happened. Woodin was able to show that if an interior model that accommodates the supercompact cardinals could be found, then, unlike the large smaller cardinals, this interior model would fit all the other major cardinals (consistent with ZFC). This remarkable result (or more precisely the work surrounding it) then gave rise to a new type of axiom, which could decide statements such as the CH. The starting point here is a return to internal models, and in particular to Gödel's $V = L$ theory, which asserts that the continuum hypothesis is true. The problem of $V = L$ is that, if there is a measurable cardinality, this equivalence is false and $V \neq L$ (Scott, 1961). Woodin, like many mathematicians, is convinced of the existence of such cardinals.

Hence, he starts by assuming that there is an internal model that can accommodate a supercompact cardinal. Note that this is a strong assumption. No such model has been found so far (in the sense that the model in question can fit a measurable cardinal). The surprising result is that, if such an internal model could be found, it would accept all major cardinals (which are consistent with ZFC).

This is a paradigm shift in all internal model design.

Where the structures produced so far by the internal model program stop somewhere near the middle of the large cardinal hierarchy (due to versions of Scott's theorem), this new model would go all the way to the top. Moreover, this internal model would be "close to V" in a well-defined sense. This makes this model, as Woodin calls it, an "Ultimate L". It would be similar to L, in the sense that it could solve undecidability problems like the continuum hypothesis. But it could at the same time accept all the major cardinal axioms (consistent with ZFC), while L does not.

7.3.3. *Ultimate L and CH*

What is "Ultimate L"? In the wake of Gödel, who had proposed, as early as 1946, a class of models definable from ordinal parameters, Woodin defined a model for ZFC named HOD (Hereditary Ordinal Definable) which, unlike L, does not have an absolute nature. If it is calculated in L, then HOD = L, but if it is calculated in V, assuming a somewhat technical assumption, then HOD $\neq L$.

It is certainly difficult, but not impossible, to study HOD. In the 1960s, Mycielski and Steinhaus, who had introduced the axiom of determination AD as an alternative to the axiom of choice, had remarked that there might be subuniverses of the universe of sets that reflect some physical intuitions which are not fulfilled by the classical sets (e.g. paradoxical decompositions of the sphere are eliminated by AD). Solovay, in 1969, had then conjectured that AD held in $L(\mathbb{R})$, relative to the existence of a supercompact cardinal. Other authors had introduced extensions or generalizations of L, in particular Mitchell and Steel who had shown, in 1994, that some could accommodate the so-called "Woodin cardinals".

$L(\mathbb{R})$ thus becomes particularly central and calls for an examination of HOD calculated on $L(\mathbb{R})$, which is written: $\text{HOD}^{L(\mathbb{R})}$. Woodin then proved the following theorem:

THEOREM 7.1 (WOODIN 2010).– *Assume there is a proper class of Woodin cardinals. Then, $HOD^{L(\mathbb{R})}$ is not a Mitchell–Steel extender model.*

This $\text{HOD}^{L(\mathbb{R})}$ belongs to a class of extender models, the most interesting of which seems to be $\text{HOD}^{L(A,\mathbb{R})}$, where A is universally Baire. If, then, relative to the extension strategy and to Woodin's Ω-Conjecture, we can name L_S^Ω the model sought, so we can introduce with Woodin the axiom $V = L_S^\Omega$. This L_S^Ω is then defined by some intersection of $\text{HOD}^{L(A,\mathbb{R})}$ and some V capable of allowing the deduction of the "essential property" sought. And thus we would have:

THEOREM 7.2 (WOODIN 2010).– *The axiom $V = L_S^\Omega$ implies the CH.*

The proof that Woodin gets the right model is still in progress, and it is not yet known if the famous "Ultimate L" will be reached. But the quest strategy is now well defined.

7.4. Infinite and philosophy

We have already addressed this question in a previous work (see Parrochia 2018, p. 218). We return to it however for the following reason: we are no longer at the time of Cantor when, the Church being still powerful, it was necessary to carefully distinguish the created (or transfinite) infinity from the uncreated (or absolute) infinity, reserved for God alone. Of course, philosophical doctrines, like physical theories indeed, today tend to dismiss the infinite from their consideration. The fact remains that certain modern philosophies – in particular that of Spinoza – by putting the arithmetic on the side of the imagination and the geometric continuum on the side of the understanding, had anticipated the countable–uncountable opposition, which Pascal knew about but Descartes or Leibniz did not. Moreover, since it placed nature (or substance or God) under the sign of the continuous (and not of the discrete), we could still wonder about the nature of the latter.

For a long time, as we know, mathematics could not answer the question of the nature of the continuum with certainty. Today, still for a moment, the continuum can be any of the alephs. Now, if we try to use mathematics to clarify certain statements of ancient philosophers, the advances on infinity since Cantor allow us to ask some questions about, for example, a philosophy like that of Spinoza.

Indeed, the Spinozist infinite can hardly be identified with one of the transfinite alephs, and there would be little sense in imagining God or nature as an inconsistent multiplicity. The fact remains, as we also observed, that contemporary philosophers have generally underestimated, if not completely misunderstood, the subtleties of recent developments in set theory, and in particular the question of large cardinals.

Of course, there is no question of asking whether God could, more or less, identify himself with a large cardinal[8]. Would it be huge, strong, super-strong, totally ineffable, indescribable or weakly inaccessible? The question is devoid of meaning, and this, even if the large cardinals of set theory, seem to have inherited the qualifiers formerly often attached to the name of God.

On the other hand, it is clear that this proliferation of large cardinals, even beyond the sequence of alephs (which already constituted an incredible transgression and posed mathematical thought as more apt to think the infinite than any other kind of discourse) consecrates the power of the human imagination and the ability of the mathematical mind to burst into reserved domains which, until then, had remained forbidden to it.

8 Moreover, the question, humorously addressed to the pope by a facetious mathematician, would risk attracting an inappropriate response of the type: "we do not have this claim". Although the papal word is reputed to be infallible, none of the men in red or black robes, edged in purple, could know such elevation: and no more the dean of the "sacred college" than the other elector or non-elector cardinals.

More than anything, the observation, at least empirical – we currently have no mathematical proof of it – of the insertion of the major cardinals in a well-ordered sequence, suggests, if we follow Woodin's conviction, that there is a sort of "natural" necessity here: this intrinsic order which assigns a very precise place to every large cardinal seems to legitimize their existence. With this hierarchy appearing as absolute as that of whole numbers, Woodin believes that the axioms of large cardinals will never lead to a contradiction. We would therefore have here an intrinsic order which would make it possible to escape the pluralism and the vagueness instituted by the theory of models and the technique of forcing.

This does not yet solve the enigma of the continuum hypothesis, but we are getting close. In 1961, the American mathematician and computer scientist Dana Scott indeed proved that if we adopt as an axiom $V = L$, i.e. if we restrict set theory to Gödel's hierarchy, then the "measurable" cardinals cannot exist. Conversely, as we have seen, Woodin is now able to prove that if we found a variant of $V = L$ compatible with the existence of a "supercompact" cardinal, then it would necessarily be compatible with all the axioms of large cardinals. The search for the "Ultimate L" not only would make forcing inoperative but would limit the impotence of mathematics based on ZFC and would make it possible to determine, for a whole class of statements – including the continuum hypothesis – whether these are true or false.

As Jean Petitot (2009, p. 130) was able to show, this evolution of mathematics is not a return to classical Platonism, once defended by a myriad of thinkers from Aristotle to Thom via Peirce, Brentano, Husserl and Weyl. Since mathematical objectivity is based on eidetic rules, we should rather, according to him, speak of a "transcendental Platonism". It is a way of seeing things, when we stick to a Kantiano–Husserlian perspective and refuse to see that the mind engages in an effective reality that it also makes it possible to transform. From a Bachelardian point of view, we would rather see here a kind of "technical noumenal Platonism", the kantian form of noumenon (the idea of the thing-in-itself if it had an existence) being relativized by Bachelard according to the power of our instruments – here our *theoretical* instruments – and becoming a *technical noumenon*.

As far as we can see, this perspective was well underlined by Jean-Paul Delahaye (2019). Since the axioms of large cardinals rank relative to each other, the fact that an axiom A is stronger than an axiom B actually means that A allows us to prove the statement asserting that "the system ZFC+ B is non-contradictory". But such a statement is expressed in theory by a sequence of symbols, so its demonstration is only a combinatorial question, which we always know how to translate into statements relating to whole numbers. The axioms of large cardinals therefore have consequences in the finite. We can even go so far as to say that the more we know about the infinite, the more we know about the finite. Moreover, these effects on arithmetic statements are, in a way, physically testable. As Delahaye writes:

> To think that a large cardinal axiom X is acceptable – and Hugh Woodin claims that many of them are – means that by making ZFC+X theory work, we are sure that we will never come across a contradiction. But to make a theory work is to unfold the consequences of its axioms, a task that can be entrusted to a machine. To believe in the existence of a large cardinal is therefore to believe that machines using the ZFC system supplemented by the axiom corresponding to this large cardinal will never produce contradictory results such as $0 = 1$ (Delahaye 2019, p. 36).

So the statement about the physical world actually comes from our understanding of infinity. In other words, advances in set theory enrich what we can say about infinity, and therefore, about the physical world. Thus, advances by mathematicians and logicians join the convictions of certain German philosophers for whom it is through the abstract that we can understand the concrete. In addition, here we will really have to wait for proof of the existence of the "Ultimate L", the culmination of this reflection in a mechanization of thought recalls Hilbert's position and leads also, of course, to a mechanistic conception of the whole world, which can just appear as an immense machinery.

8

In Search for a New Philosophy

Assume that the world is a graph. There are two cases to be examined: either this graph is finite or it is infinite. In the following, we will explore the two possibilities, knowing that, in each case, the philosopher's quest is that of a stable subset whose name has always been called "the Absolute".

8.1. The finite case

Let W be the world and $W(X, U)$ the finite graph associated with W. The question is: on what condition is W or some $G \subset W$ stable?

Here, we will appeal to the Erdös–Hajnal conjecture, which states that families of graphs defined by forbidden induced subgraphs have either large cliques or large independent (i.e. stable) sets.

Recall that in graph theory, an independent set (or stable set) is a set of vertices in a graph, where no two vertices are adjacent. See, for example, the independent set for the general Petersen graph GP(12,4) in Figure 8.1.

More precisely, we know that, for an arbitrary undirected graph H, we call $\mathcal{F}_H$ the family of graphs that do not have H as an induced subgraph. Then, according to the conjecture, there exists a constant $\delta_H > 0$, such that the n-vertex graphs in $\mathcal{F}_H$ have either a clique or an independent set of size $\Omega(n^{\delta_H})$.

An equivalent statement to the original conjecture is that, for every graph H, the H-free graphs all contain polynomially large perfect induced subgraphs.

So, if we want to identify the Absolute with a relatively large stable set, we will instead choose graphs that satisfy the Erdös–Hajnal conjecture.

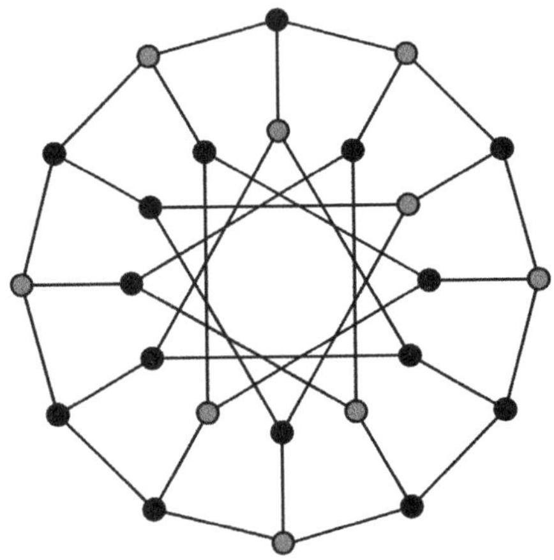

Figure 8.1. *A maximal independent set (source: Wikipedia article: Independent set (graph theory))*

This conjecture is proved to be true for bull graphs[1] and, more generally, when H is a cograph. The authors also showed that, for arbitrary H, the size of the largest clique or independent set grows superlogarithmically. More precisely, for every H, there is a constant c, such that the n-vertex H-free graphs have cliques or independent sets containing at least $\exp c\sqrt{\log n}$ vertices. The graphs H for which the conjecture is also true include the four-vertex path graph (see Chudnovsky 2014; Gyárfás 1997), the five-vertex bull graph (see Chudnovsky 2014; Nadis 2021) and any graph that can be obtained from these and the cographs by modular decomposition (see Chudnovsky 2014; Alon et al. 2001). As of 2014, however, the full conjecture has not been proven, and remains an open problem (see Chudnovsky 2014).

1 The bull graph is a planar undirected graph with five vertices and five edges, in the form of a triangle with two disjoint pendant edges (see Chudnovsky 2014). It has chromatic number 3, chromatic index 3, radius 2, diameter 3 and girth 3. It is also a self-complementary graph, a block graph, a split graph, an interval graph, a claw-free graph, a one-vertex-connected graph and a one-edge-connected graph. A graph is bull-free if it has no bull as an induced subgraph. The triangle-free graphs are bull-free graphs, since every bull contains a triangle. The strong perfect graph theorem was proven for bull-free graphs long before its proof for general graphs (see Alon et al. 2001), and a polynomial time recognition algorithm for bull-free perfect graphs is known (see Erdös and Rényi 1963). Chudnovsky and Safra studied bull-free graphs more generally, showing that any such graph must have either a large clique or a large independent set, and developed a general structure theory for these graphs (see Chudnovsky and Safra 2008; Chudnovsky et al. 2021; Nadis 2021).

An earlier formulation of the conjecture, also by Erdös and Hajnal, and still unsolved, concerns the special case when H is a five-vertex cycle graph (see Gyárfás 1997). The C_5-free graphs include the perfect graphs, which necessarily have either a clique or independent set of size proportional to the square root of their number of vertices. Conversely, every clique or independent set is itself perfect. For this reason, a convenient and symmetric reformulation of the Erdös–Hajnal conjecture is that for every graph H, the H-free graphs necessarily contain an induced perfect subgraph of polynomial size.

Finding a maximum independent set can be tricky: the problem is deemed NP-hard. Then, as we know, the philosophers of the Classical Age, like the logicians of the modern era, have rather sought to reduce the size of the set of statements on which their constructions are based, in other words the elements that can be precisely identified with the independent or stable set of a graph. On this basis, we may think that there is no need to search for a maximal independent set of minimal size. Indeed, the size of a maximal independent set can vary greatly. For example, the following two independent sets for the star graph S_8 show how vastly different in size two maximal independent sets (the right being maximum) can be.

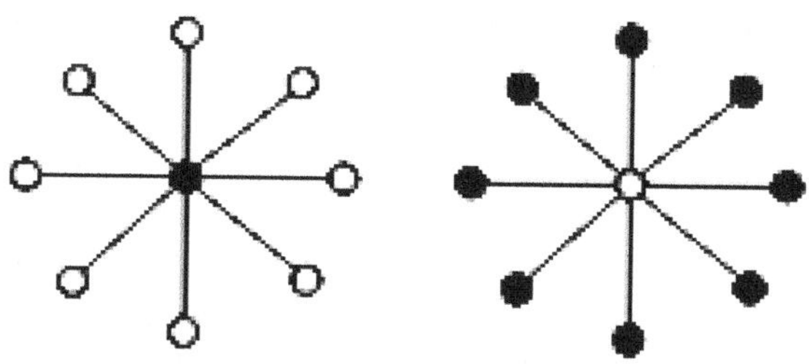

Figure 8.2. *Two maximal independent sets for S_8*

Another result may be used.

Let us introduce the following notations: the vertex set of a graph G is $V(G)$, the order of G is $n = n(G) = |V(G)|$, the set of neighbors of a vertex v is $N_G(v)$ (or simply $N(v)$ if there is no possibility of ambiguity), the set of neighbors of a set $S \subseteq V(G)$ in G is $N_G(S) = \bigcup_{u \in S} N(u)$ (or simply $N(S)$ if there is no possibility of ambiguity), the graph induced on S is $G[S]$ and the independence number, the cardinality of a maximum independence set (MIS), is $\alpha(G)$. All graphs are assumed to be finite and simple.

DEFINITION 8.1.– *An independent set of vertices* I_c *is a critical independent set if* $|I_c| - |N(I_c)| \geq |J| - |N(J)|$, *for any independent set* J.

A graph may contain critical independent sets of different cardinalities:

– a graph consisting of a single edge (K_2, the complete graph on two vertices) has critical independent sets of cardinalities 0 and 1;

– for some graphs, the only critical independent set is the empty set; K_3 is an example.

DEFINITION 8.2.– *A maximum critical independent set is a critical independent set of maximum cardinality.*

It is easy to verify that, for any graph with at least three vertices, a maximum critical independent set must contain all pendant vertices. Therefore, a maximum critical independent set is a generalization of the set of pendants. The critical independence number of a graph G, denoted as $\alpha' = \alpha'(G)$, is the cardinality of a maximum critical independent set. If I_c is a maximum critical independent set, and therefore $\alpha'(G) = |I_c|$, then clearly $\alpha' \leq \alpha$.

DEFINITION 8.3.– *The critical independence number of a graph is the cardinality of a maximum critical independent set.*

The critical independence number of a graph is a lower bound for the independence number and can be computed in polynomial time. A result from Larson (2011) proves that any graph can be efficiently decomposed into two subgraphs where the independence number of one subgraph equals its critical independence number, where the critical independence number of the other subgraph is zero, and where the sum of the independence numbers of the subgraphs is the independence number of the graph.

So, while directly finding that an independent set in a graph is a NP-hard problem, this decomposition allows us to find a solution with an efficient algorithm in a polynomial time. Now, according to Butenko and Trukhanov, if I_c is a critical independent set in a graph G, then there is a maximum independent set I in G, such that $I_c \subseteq I$.

Bipartite graphs are independence reducible, but some non-bipartite graphs also have this property. In contrast, some graphs are independent irreducible. In them, for any non-empty independent set of vertices I, we always have $|N(I)| > |I|$. We can see some examples of this in Figure 8.3.

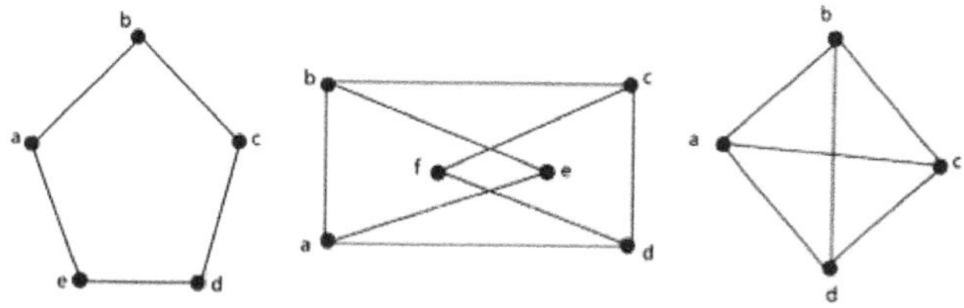

Figure 8.3. *Independent irreducible graphs for any non-empty independent set of vertices*

Philosophers who seek to reduce the graph of their philosophical architectonics to an independent set should therefore avoid this kind of configuration, except if they wanted, in the Hegelian way, to identify the whole of their system with the absolute itself.

But now we want more. The philosopher who would like to extract from the asymmetric graph of the world a subgraph consisting of an independent set of vertices would have an interest in applying the preceding theorems to avoid being faced with an NP-hard problem.

The problem of such an extraction is indeed fascinating and comes in different stages.

1) How, among the multiple elements of the world, can we find a subset that is, without bias, representative of the whole?

2) How do we find in it the maximum independent (or stable) subset, the underlying skeleton that explains everything else?

A variant of operation (2) could be:

3) How do we find the maximal tree of the graph that summarizes it while keeping only the essential links?

These concentration-reduction operations are not trivial. To be carried out, they would require a knowledge which exceeds that which a human being can reasonably acquire during their lifetime. Consequently, they should mobilize either several men or a technical apparatus capable of exploring all the known aspects of the world to extract the substantive marrow.

The philosophers of the past, of course, never did so.

Starting from a precise problem (problem of the one and the multiple in the pre-Socratics, problem of truth in Plato, problem of the foundation of science in Descartes, problem of the possibility of metaphysics as science in Kant, problem of contradiction between the absolute and time in Hegel, problem of the contradiction between evolution and creation in Bergson, etc.), they tried to answer it by developing a theory based on particular concepts (pre-Socratic elements, Platonic idea, cogito and God in Descartes, noumenon–phenomenon distinction in Kant, idea of a logic of history in Hegel, concepts of duration and creative evolution in Bergson, etc.). Therefore, many fixed points are supposed to explain reality without having to travel through the world of beings.

Worse: from Nietzsche and Freud, the entry into the era of suspicion has rather consisted of supposing, under appearances, hidden intentions or susceptible archaeological foundations, in order to explain the very movement of ideas. Instead of investigating the world itself, philosophy, abandoning experience, was instead constituted – at least in part – by application to itself, becoming, in a way, a "meta-philosophy", but in fact developed without scientific rigor.

Phenomenology, of course, has attempted to return to things, but not to objects. Suspending all knowledge, it got lost in the painting of a world experienced by a subject in a supposedly primitive experience preceding predicative judgments. But apart from the fact that the supposedly primitive is not one – it is always contaminated by history, including in the language which tries to express it – there is no reason for it to be held for fundamental, but rather as that which must necessarily be transcended. But if we now have to take into account science, technology and their history, the task becomes almost insurmountable for a human brain.

If they want to continue their task, the philosopher must then necessarily use the theoretical and concrete tools which alone can give them access to the immense reserves of knowledge accumulated in libraries, laboratories and all the places that hoard them from the world. They will then have to take not only their own culture, but also others, into account.

8.2. The infinite case

Suppose that the world (or what the philosopher interprets as such, i.e. what they reduce it to) is an infinite graph. What does it mean to look for a stable set in an infinite graph?

Let us first specify the notion of an infinite graph, which is not univocal. For this, it is best to use the notations of Nash-Williams Nash-Williams (1968).

DEFINITION 8.4.– *The set of vertices of a graph G will be denoted, as usual, by $V(G)$ and its set of edges will be denoted by $E(G)$. G is finite if $|V(G) \cup E(G)|$ is finite, enumerable if $|V(G) \cup E(G)| = \aleph_0$, countable if it is finite or enumerable and locally finite if the degree of every vertex is finite (i.e. if each vertex is incident with only finitely many edges).*

Let us now introduce some new definitions that generalize some preceding ones.

DEFINITION 8.5 (WALK).– *A walk in G is a finite, one-way infinite or two-way infinite sequence of one of the following forms:*

$$v_0, e_1, v_1, e_2, v_3, e_3, ..., e_n, v_n \quad \textit{(finite walk)}$$

or:

$$v_0, e_1, v_1, e_2, v_3, e_3, ... \quad \text{(one-way } \textit{infinite walk)}$$

or:

$$..., e_{-2}, v_{-2}, e_{-1}, v_{-1}, e_0, v_3, v_0, e_1, v_1, e_2, ... \quad \text{(two-way } \textit{infinite walk)}$$

where the v_i are vertices of G and each e_i is an edge joining the vertices immediately preceding and following it in the sequence.

DEFINITION 8.6 (TRAIL, PATH).– *A walk is a trail if no edge appears more than once in it, and is a path if no vertex appears more than once.*

DEFINITION 8.7 (CLOSED WALK, CIRCUITS).– *A finite walk is closed if its first and last terms are the same. A closed walk with more than one term in which there are no repetitions apart from the first and last terms being the same is a circuit.*

DEFINITION 8.8 (EULERIAN AND HAMILTONIAN CIRCUITS).– *A trail in G is Eulerian if every vertex and edge of G appears in it. A path or circuit in G is Hamiltonian if every vertex of G appears in it.*

DEFINITION 8.9 (DECOMPOSITION).– *A set of trails in G constitute a decomposition of G if every vertex of G appears in at least one of them and every edge of G appears in exactly one of them (so that the trails subdivide $E(G)$ into disjoint subsets).*

Also recall that two vertices of a graph are adjacent if they are joined by an edge.

The study of infinite graphs begins with the famous König's lemma (see König 1990, p. 13):

THEOREM 8.1 (KÖNIG'S LEMMA).– *Let $S_1, S_2, S_3, \ldots$ be an infinite sequence of disjoint non-empty finite sets and $\prec$ be a relation in $S_1 \cup S_2 \cup \ldots$ such that whenever n is a positive integer and $x \in S_{n+l}$, there exists a $y \in S_n$ such that $y \prec x$. Then, there exists an infinite sequence $x_1, x_2, x_3, \ldots$ such that $x_n \in S_n, (n = 1, 2, \ldots)$ and $x_1 \prec x_2 \prec x_3 \prec \ldots$.*

An illustrative application is as follows:

THEOREM 8.2.– *Let k be a positive integer and G be an enumerable graph. If every finite subgraph of G is k-colorable, then so is G (we call G k-colorable if its vertices can be colored using at most k colors so that no two adjacent vertices have the same color).*

In this case, we get this trivial consequence:

THEOREM 8.3.– *G admits at least one independent (stable) subset.*

The following theorem, which is Lemma 1 of Rado (1949) and Theorem 7.1.3 of Ore (1962), can be used to prove that a general (not necessarily enumerable) infinite graph is K-colorable if all of its finite subgraphs are k-colorable, with k being a positive integer.

THEOREM 8.4.– *Let Y, Z be sets and let $\mathfrak{F}$ denote the class of all finite subsets of Y. For each $A \in \mathfrak{F}_Y$, let f_A be a function from A into Z and, for each $y \in Y$, let the set $\{f_A(y) : y \in A \in \mathfrak{F}_Y\}$ be finite. Then, there exists a function $f : Y \to Z$ such that, for every $A \in \mathfrak{F}_Y$, there exists a $B \in \mathfrak{F}_Y$ such that $A \cup B$ and $f|A = f_B|A$ (the notation $f|A$ indicates the function $g : A \to Z$ such that $g(y) = f(y)$ for every $y \in A$, and $f_B|A$ is similarly defined).*

To prove that an infinite (not necessarily enumerable) graph G is k-colorable if all its finite subgraphs are k-colorable, let the set Y of Theorem 8.2.4 be $V(G)$ and let $Z = \{C_1, \ldots, C_k\}$ be a set of k colors. If $A \in \mathfrak{F}_Y$, the k-colorability of every finite subgraph of G ensures that we can select a function $f_A : A \to Z$, such that no two elements of A, which are adjacent in G, have the same image under f_A. Assuming the Axiom of Choice, we thus select an f_A for every $A \in \mathfrak{F}_Y$, and then it is not difficult to see that a function $f : Y \to Z$ satisfying the condition of Theorem 8.2.4 gives a coloring of G using at most the colors $\{C_1, \ldots, C_k\}$, such that no two adjacent vertices of G receive the same color. This theorem concerning k-colorability of infinite graphs was first established by Erdös and De Bruijn (1951).

Theorem 8.2.4 is useful, in suitable cases, for extending known theorems concerning finite graphs to infinite graphs that are not necessarily enumerable.

A trivial consequence here is as follows:

THEOREM 8.5.– *Let G be a non-countable graph. G admits at least one independent (stable) subset.*

The philosophical consequence of the previous theorem is that philosophical systems using infinite structures (see, for example, Spinoza's philosophy) can perfectly exhibit infinite independent (or stable) sets. Infinite attributes or infinite modes are their examples in Spinoza's *Ethics*.

In conclusion, supposing that philosophers reduce their pretensions to the search, not exactly of the Absolute, but only of some kind of invariant, such as an independent set, they can continue to find elements of stability in the world, behind the chaotic play of appearances.

9

Extension of Structuralism and Negative Theology

Assume again that the structure of the world is a graph in the sense of Dipert (1997), i.e. it is some kind of finite asymmetric graph G. Suppose now this graph has some additional property: for example, we have defined on it an ordered structure (chain) or a partially ordered structure (poset, lattice) or a generalization of this one (matroid, simplicial complex). Imagine then that you want to study not only the explicit relations between the elements of G but also the implicit ones, i.e. those which result from the complementary graph $\overline{G}$ of G.

This request engenders two consequences:

1) we obtain a fantastic extension of structuralism which, in the social sciences, has hitherto been limited to describing – in all the disciplines where it has been applied (linguistics, ethnology, etc.) – explicit structures;

2) we thus encounter a problematic which, in the past, was that of negative theology: let us recall that, unable to say what God is (if God is transcendent, he is absolutely different and beyond the reach of human understanding), a number of authors, since antiquity, have decided that we cannot say what God is but only what He is not. They argued then that, by going to the complementary, they still could approach, in a certain way, what God is. The method has its model in set theory, where the complement of the complement of a set is nothing but the set: $C(C(A)) = A$. However, we will see that the theory of graphs and the theory of orders limit the consequences that can be obtained by such a project.

Knowing what a graph is, especially a simple graph, with no loops, let us now recall some of the definitions already introduced in Chapter 1 by illustrating some examples.

DEFINITION 9.1 (ADJACENCY MATRIX OF A GRAPH).– *We recall that the adjacency matrix of a graph is a matrix* $M_{i,j}(i = 1, ..., n; j = 1, ..., n)$ *with coefficients 0 or 1, where each row corresponds to a vertex of* G*, where each column corresponds to a vertex of G and where:*

$$M_{ij} = \begin{cases} 1 & \text{if and only if } (i,j) \in U; \\ 0 & \text{otherwise.} \end{cases}$$

In this chapter, we are dealing with undirected simple graphs.

Among the graphs, some present regularities: for example, the same "degree", i.e. the same number of edges or arcs pertaining to each vertex. They are called "regular" graphs. Others are not like this: they are "irregular", or even "highly irregular", and are today the object of numerous studies and of many conjectures (see, for example, Chartland (2016, pp. 1–16)).

We will leave aside this question here to focus on complementarity graphs introduced in Chapter 1 (a question often neglected in graph theory[1]). We will go on by studying the comparability and incomparability graphs on some ordered structures.

9.1. Complementarity graphs

DEFINITION 9.2 (COMPLEMENTARY GRAPH OF A SIMPLE GRAPH: SEE GONDRAN AND MINOUX (1979), P. 4).– *Given a simple graph* $G = [X, U]$*, the complementary graph* $G' = [X, \overline{U}]$ *has the same set of vertices as G and as edges, the complementary edges of those of* U*. We have:*

$$(i,j) \in U \Rightarrow (i,j) \notin \overline{U};$$

$$(i,j) \notin U \Rightarrow (i,j) \in \overline{U}.$$

Figure 9.1 shows a simple graph G and its complementary graph $\overline{G}$.

As we can see, the complementary $\overline{G}$ of the graph G is disconnected. In general, as we know, the complementary graph of a disconnected graph is connected, but the converse is not true.

1 We have to note that certain books, although reputed to be "bibles" of graph theory, devoted very few lines to these questions, and in particular to comparability and incomparability graphs, the properties of which are often left to prove to the reader (see Diestel 2001, 127, 136 and 242).

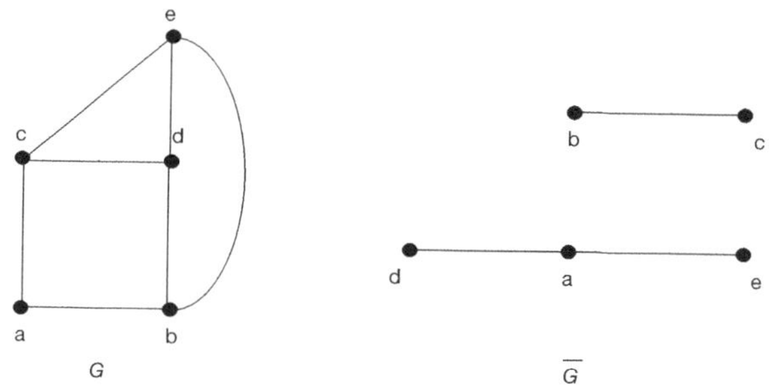

Figure 9.1. *Graph G and its complementary $\overline{G}$*

DEFINITION 9.3 (COMPLEMENTARY MATRIX OF A SIMPLE GRAPH).– *We call complementary matrix of a simple graph G the matrix M' associated with the complementary graph $\overline{G}$ of G.*

THEOREM 9.1.– *The complementary matrix is obtained as the difference between the inverse matrix $\overline{M}$ of the adjacency matrix of the graph G and the unit matrix I_n if G has n vertices.*

PROOF.– Consider $\overline{M}$, the matrix where the 1s of M have been changed into 0s, and conversely, the 0s have been changed into 1s. A consequence is necessarily that $\overline{M}$ is a matrix of a loop graph because the 0s of the diagonal of M have been replaced by 1s, which means that a is in relation to itself, b in relation to itself, etc. If we want to obtain as complement of G a simple graph $\overline{G}$ without loops, we need to find a matrix M' with zero diagonal. In other words, we must subtract 1s from 1s on the diagonal of M and, for that, we must ask:

$$M' = \overline{M} - I_n,$$

where I_n is the identity matrix with n rows and n columns, whose diagonal terms are all equal to 1 and non-diagonal terms equal to 0. □

Knowing the number of edges of $G[X, U]$, it is easy to calculate the number of edges of its complementary graph.

THEOREM 9.2.– *If* $|X| = n$, $\overline{|U|} = \binom{n}{2} - |U|$.

PROOF.– To each edge of a graph corresponds a pair of vertices. So, the total number of all possible edges of a graph is the total number of all possible combinations of two elements taken from n, say $\binom{n}{2} = N$. To get the cardinal of $\overline{U}$, it suffices to subtract from this number N the cardinal $|U|$ of the set of edges of $G[X, U]$ since $|U| + \overline{|U|} = N = \binom{n}{2}$. □

9.2. Order relation, ordered set

DEFINITION 9.4 (ORDER RELATION – RECALL FROM 3.1 AND KAUFMANN AND BOULAYE (1978), PP. 5–7).– *A binary internal relation $\leq$ on a non-empty set E is an order relation if for all x, y and z elements of E, we have:*

1) $x \leq x$ (reflexivity);

2) $(x \leq y$ and $y \leq x) \Rightarrow x = y$ (antisymmetry);

3) $(x \leq y$ and $y \leq z) \Rightarrow x \leq z$ (transitivity).

If we replace $\leq$ by $<$, then the relation becomes a "strict order" relation. In this case, it is irreflexive. It reads "x is less than y", or "y is greater than x" and we have:

$x < y$ if and only if $x \leq y$ and $x \neq y$.

Recall too that some order relations are *total*, i.e. two elements of the set E are always *comparable*, in other words, for all x, y of E:

$x \leq y$ or $y \leq x$.

We then say that $\leq$ is a relation of "total order" and that the set E is *totally ordered* by this relation.

DEFINITION 9.5 (ORDERED SET – SEE ALSO 3.1 AND HARZHEIM (2005)).– *An ordered set is a set provided with an order relation.*

DEFINITION 9.6 (PARTIAL ORDER – PARTIALLY ORDERED SET (OR POSET)).– *An order relation on E is said to be "partial" if it is not total, and E is then a partially ordered set (or poset).*

DEFINITION 9.7 (NEGATION OF AN ORDER RELATION).– *The negation of a binary relation R defined on a set E is the relation associated with the complementary graph of the relation R in $E \times E$. It is usually written $\not R$. In other words, two elements are related by $\not R$ if and only if they are not related by R.*

To say that an order is total is to say that its negation is the strict reverse order. This means that there is an equivalence for an order $\leqslant$ between:

– $\leqslant$ is total;

– $x \nleqslant y \iff y < x$.

On the other hand, as soon as there are two distinct elements not comparable within an order, the negation of the order relation cannot be an order (strict or not) because it is not antisymmetric. The negation of a non-total order is therefore never an order.

9.3. Graphs associated with a partially ordered set

DEFINITION 9.8 (COMPARABLE AND INCOMPARABLE ELEMENTS).– *Let P be a partially ordered set, endowed with the relation $<$, reflexive, antisymmetric, transitive. We say that two elements x and y of P are comparable if $x < y$ or $y < x$ (we can also replace $<$ by $\leq$). Otherwise, i.e. when two elements of P are incomparable, we have between them a relation $||$ called "relation of incomparability", reflexive, symmetric and transitive.*

DEFINITION 9.9 (COMPARABILITY AND INCOMPARABILITY GRAPHS).– *When P is a partially ordered set (poset), we can thus associate with P two graphs. One is called comparability graph of P and the other incomparability graph of P. These two graphs have the elements of P as a set of vertices. In the comparability graph, two distinct elements x and y of P are adjacent when they are comparable in P. Likewise, x and y are adjacent in the incomparability graph when they are incomparable in P.*

We will focus more specifically on the latter case. Before that, let us introduce the following definitions:

DEFINITION 9.10 (CLIQUE).– *A clique C in an undirected simple graph $G = (X, U)$ is a subset of the vertices, $C \subseteq X$, such that every two distinct vertices are adjacent. This is equivalent to the condition that the subgraph of G induced by C is a complete graph.*

DEFINITION 9.11 (COLORING A GRAPH).– *The coloring of the vertices (resp. edges) of a graph $G = (X, U)$ corresponds to the assignment of a color to each of its vertices (resp. edges) so that two adjacent vertices (respectively, edges) do not carry the same color.*

DEFINITION 9.12 (CHROMATIC NUMBER OF A GRAPH).– *A graph is said to be p-chromatic if its vertices admit coloring in p colors. We call chromatic number $\chi(G)$ (respectively, chromatic index $q(G)$) the minimum number of distinct colors needed to color the vertices (respectively, edges) of the graph G.*

DEFINITION 9.13 (PERFECT GRAPH – SEE BERGE AND CHVÁTAL (1984)).– *A perfect graph is a graph where the chromatic number of every induced subgraph equals the size of the largest clique of that subgraph (i.e the clique that has the maximum number of vertices Ð or "clique number"). Equivalently, stated in symbolic terms, an arbitrary graph $G = (X, U)$ is perfect if and only if:*

$$\text{for all } C \subseteq X \text{ , we have } \chi(G[C]) = \omega(G[C]).$$

In 1950, Dilworth (1950) was able to prove that incomparability graphs were perfect graphs, so that the chromatic number of an incomparability graph is equal to its number of cliques (an analogous result for the comparability graphs had previously been proven by Erdös and Szekeres (1935)). Gallai (1967) gave a characterization of incomparability graphs in terms of minimal induced prohibited subgraphs. We also know that the incomparability graphs can be recognized in a polynomial time (see Golumbic 1980). In 1983, Golumbic, Rotem and Urrutia (see Golumbic et al. 1983) and Lov'asz (1983) proved that any incomparability graph was also a string graph (on all of this, see Fox and Pach (2012)). But this does not allow us to precise the structure of these graphs. We will show that, in some cases, we can directly compute the number of edges of their incomparability associated graphs.

It will be found in Figure 9.2 an example of poset (A), its comparability graph (B) and its incomparability graph (C).

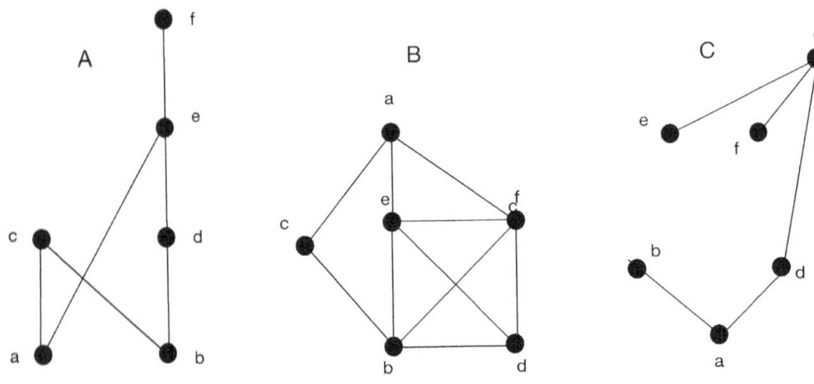

Figure 9.2. *Poset, comparability graph, incomparability graph*

9.4. Complementarity and incomparability graphs of a poset

Let P be a poset and $D(X,U)$ be the Hasse diagram of P. The set of edges U is formed by the explicit order relations in P, i.e.:

$$U = \{(x,y) \in X \mid x \leq y\}.$$

Now let $G(P)$ be the comparability graph of the poset P. Let E be the set of edges of $G(P)$. E is formed by all the ordered pairs of elements of P, and we have:

$$E = \{(x,y) \in P \mid x \leq y\}.$$

Hence, we can deduce that $E - U$ is the set of non-explicit order relations between the elements of P.

Now let $D(X,\overline{U})$ be the complementary graph of $D(X,U)$, the Hasse diagram of P. We can now deduce the following proposition:

THEOREM 9.3.– *The number N of edges of $I(P)$, the incomparability graph of P, is given by the difference between the cardinal of $\overline{U}$ and the cardinal of the set of non-explicit order relations between the elements of P. We have:*

$$N = |\overline{U}| - |E - U| = |\overline{U}| + |U| - |E| = \binom{n}{2} - |E|.$$

In other words, knowing the number of edges of the comparability graph of a poset P, we can easily get the number of edges of its incomparability graph. And, of course, conversely, knowing the number of edges N of the incomparability graph of a poset P, we can easily get the number of edges of its comparability graph, since $|E| = \binom{n}{2} - N$.

Examples

In the case of the poset of Figure 9.2A, we have $U = 6$, $|\overline{U}| = 9$, $|E| = 10$. Hence, $|E - U| = 10 - 6 = 4$ and $N = |\overline{U}| - |E - U| = 9 - 4 = 5$. We obtain this result more quickly by simply writing $N = \binom{6}{2} - 10 = 15 - 10 = 5$.

9.5. Boolean representation of a poset

In fact, we can calculate in a very simple way the number of edges of a poset incomparability graph. There are indeed Boolean representations of posets (which can also be applied to lattices, simplicial complexes and matroids (see Rhodes and

Silva, 2015))[2]. We just define the Boolean matrix $M(P) = m_{xu}$ of a poset P by setting:

$$m_{xu} = \begin{cases} 0 & \text{if } x \geq u; \\ 1 & \text{otherwise.} \end{cases}$$

Then, we have just to compute in the matrix the 0s which are not on the diagonal. By this method, we directly get the cardinal of the comparability graph of P. Hence, we can easily deduce the cardinal of its incomparability graph. In the case of the poset of Figure 9.2, the matrix is given in Table 9.1:

$m_{xu} =$

	a	b	c	d	e	f
a	0	1	1	1	1	1
b	1	0	1	1	1	1
c	0	0	0	1	1	1
d	1	0	1	0	1	1
e	0	0	1	0	0	1
f	0	0	1	0	0	0

Table 9.1. *Comparability matrix of* P

Of course, we cannot permute the 0s and the 1s of this matrix to get the matrix of the incomparability graph: this one has not the same number of vertices. Besides, it would make no sense to do so. The following example will suffice to show it: Let $T = \{0, a, b, 1\}$ with $0 \leq a, b \leq 1$. Since we do not have $a \geq b$ (as a and b are incomparable), $m_{ab} = m_{ba} = 1$. But by passing to the matrix M_c complementary to M, we necessarily obtain $m_{ab} = m_{ba} = 0$, which means that we have both $a \geq b$ and $b \geq a$. But this results in $a = b$, which is wrong. So, a problem could be: how do we directly get the number of edges of an incomparability graph $I(P)$, without knowing the number of edges of the comparability graph of the poset P? We will see that, in the case of Boolean lattices, this is perfectly possible.

2 There is an alternative possibility: following Stanley (1970, 1988, 2004) and more recent researchers (see Wu and Shizen 2012), we shall say that the incidence matrix $\mathfrak{m}_P$, which encodes full information about the poset P, is the indicator function of $<$ and may be described as follows:

$$\mathfrak{m}_P = \begin{cases} 1 & \text{if } x < y; \\ 0 & \text{otherwise.} \end{cases}$$

9.6. Case of lattices

As we know (see Birkhoff 1940), a lattice $(T, \leq)$ is a partially ordered set (poset) in which every element has one and the same supremum (least upper bound) and one and the same infimum (greatest lower bound). Figure 9.3 gives a typical example.

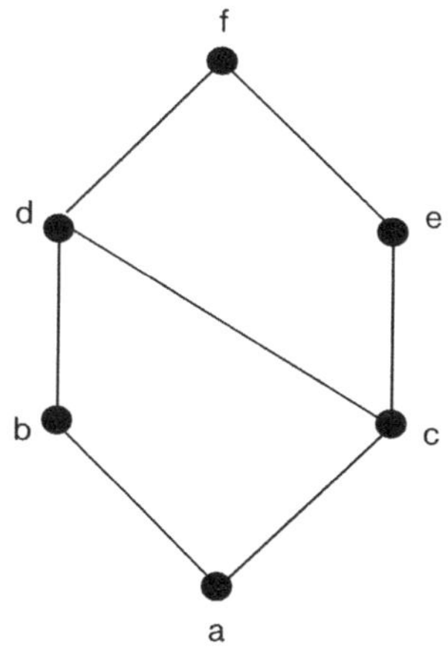

Figure 9.3. *A classic example of a lattice*

Suppose we want to know explicitly the matrix of the incomparability graph of this lattice and how it is connected to the matrix of its complementary graph. We could proceed as follows: draw the adjacency matrix of the Hasse diagram of the lattice, then the matrix of its complementarity graph. The matrix of the incomparability graph can be deduced very easily.

The adjacency matrix $M(T)$ of the Hasse diagram of T, the lattice of Figure 9.3, is given in Table 9.2.

M (T) =

	a	b	c	d	e	f
a	0	1	1	0	0	0
b	1	0	0	1	0	0
c	1	0	0	1	1	0
d	0	1	1	0	0	1
e	0	0	1	0	0	1
f	0	0	0	1	1	0

Table 9.2. *Adjacency matrix of* T

The matrix $M_c(T)$, complementary to $M(T)$, appears in Table 9.3.

$M_c(T) =$

	a	b	c	d	e	f
a	0	0	0	1	1	1
b	0	0	1	0	1	1
c	0	1	0	0	0	1
d	1	0	0	0	1	0
e	1	1	0	1	0	0
f	1	1	1	0	0	0

Table 9.3. *Adjacency matrix of the complement of T*

Figure 9.4 shows the complementary graph (A) and the incomparability graph (B) of the lattice T. As we can see, the incomparability graph of T has six vertices and only three edges.

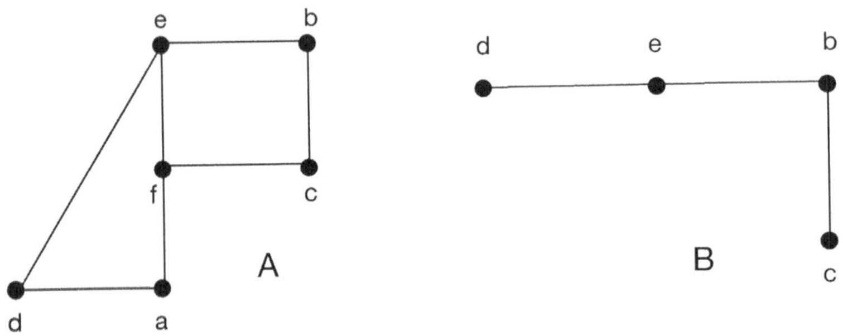

Figure 9.4. *Complementary graph (A) and incomparability graph (B) of the Hasse diagram of the lattice T*

Table 9.4 gives the matrix $M'(T)$ of the incomparability graph of T:

M'(T) =

	b	c	d	e
b	0	1	0	1
c	1	0	0	0
d	0	0	0	1
e	1	0	1	0

Table 9.4. *Adjacency matrix of the incomparability graph of T*

Note that $M'(T)$ can be obtained from $M_c(T)$ by deleting the first and last rows, and the first and last columns.

We may observe that two non-isomorphic lattices can have the same incomparability graph (see Figure 9.5).

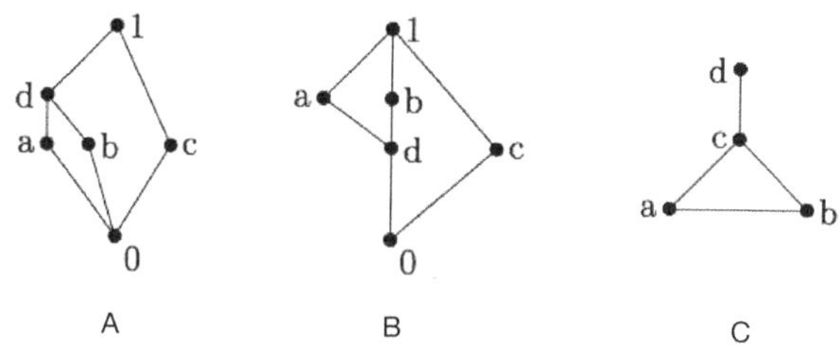

Figure 9.5. *Non-isomorphic A and B lattices having the same incomparability graph C*

As Wasadikar and Survase (2012) point out, there are many papers that interlink lattice theory and graph theory. Filipov (1980) was undoubtedly one of the first to study the comparability graphs of partially ordered sets by defining the adjacency between two elements of a poset by the comparability relation: a and b are adjacent if $a \leq b$ or $b \leq a$. In the same order of ideas, Duffus and Rival (1977) define the overlap graph of a poset by the overlap relation: a and b are adjacent if $a \prec b$ or $b \prec a$. Articles by Gedenova (1980) and Bollobás and Rival (1979) discuss the properties of overlap graphs derived from lattices. In Nimbhorkar et al. (2010), Nimbhorkar, Wasadikar and Pawar introduced graphs on a lattice L with 0, and defined adjacency of two elements $x, y \in L$ by the relation: $x \wedge y = 0$. More recently, Bostjan Bresar et al. (2008) have themselves introduced cover graphs which are incomparability graphs of posets. In particular, they defined a graph in which the set of edges is the union of the set of edges of the cover graph and of the corresponding incomparability graph. In short, a lot of works have focused on these questions since the 1980s.

In their paper, Wasadikar and Survase draw all graphs with a number of edges $n \leq 4$ which are incomparability graphs associated with lattices and give some simple properties of these graphs without relating them to particular lattices. On the contrary, we will show that certain types of lattices make it possible to define additional properties, in particular Boolean lattices, which make it possible to compute directly the number of edges of the incomparability graph associated with them.

9.6.1. *Case of Boolean lattices*

As we know, Boolean structures (in conjunction with order theory) are of particular importance (see Ponasse and Carréga 1979; Rhodes and Silva 2015) in many fields. The Boolean lattices T_n are of the form $\{0,1\}^n$. Consider, for example, the lattice T_3, isomorphic to $\{0,1\}^3$, shown in Figure 9.6.

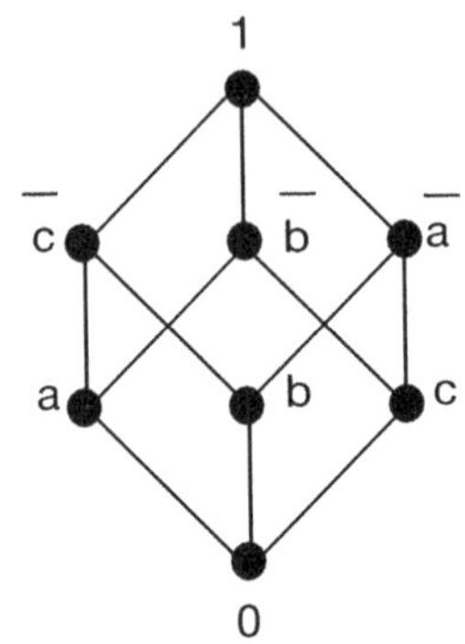

Figure 9.6. *Hasse diagram of* T_3

The adjacency matrix of the associated graph is that of Table 9.5.

M(T_3) =

	0	a	c	$\bar{b}$	b	$\bar{c}$	$\bar{a}$	1
0	0	1	1	0	1	0	0	0
a	1	0	0	1	0	1	0	0
c	1	0	0	1	0	0	1	0
$\bar{b}$	0	1	1	0	0	0	0	1
b	1	0	0	0	0	1	1	0
$\bar{c}$	0	1	0	0	1	0	0	1
$\bar{a}$	0	0	1	0	1	0	0	1
1	0	0	0	1	0	1	1	0

Table 9.5. *Adjacency matrix of* T_3

The complementarity matrix of the graph associated with T_3 is represented in Table 9.6.

Hence the associated graph, which can take several configurations (see Figure 9.7).

We can go to the incomparability graph as before, by removing the vertices which are in an order relation (in this case, 0 and 1), as well as the related edges. The graph being regular of degree 4, there would in principle be 8 edges to remove, but 0 and 1 being connected by an order relation, they have one edge in common. We must therefore remove 8 - 1 = 7 edges. Since the complementary graph of T_3 has 8 vertices and 16 edges, the incomparability graph will therefore have 6 vertices and 16 - 7 = 9 edges (see Figure 9.8).

$$M_c(T_3) = \begin{array}{c|cccccccc} & 0 & a & c & \bar{b} & b & \bar{c} & \bar{a} & 1 \\ \hline 0 & 0 & 0 & 0 & 1 & 0 & 1 & 1 & 1 \\ a & 0 & 0 & 1 & 0 & 1 & 0 & 1 & 1 \\ c & 0 & 1 & 0 & 0 & 1 & 1 & 0 & 1 \\ \bar{b} & 1 & 0 & 0 & 0 & 1 & 1 & 1 & 0 \\ b & 0 & 1 & 1 & 1 & 0 & 0 & 0 & 1 \\ \bar{c} & 1 & 0 & 1 & 1 & 0 & 0 & 1 & 0 \\ \bar{a} & 1 & 1 & 0 & 1 & 0 & 1 & 0 & 0 \\ 1 & 1 & 1 & 1 & 0 & 1 & 0 & 0 & 0 \end{array}$$

Table 9.6. *Adjacency matrix of the complementary graph of T_3*

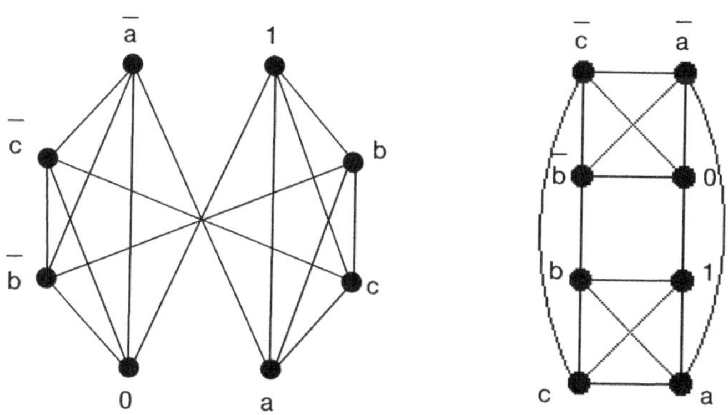

Figure 9.7. *Complementary graph of T_3*

At the matrix level, this amounts to removing the first and the last row, as well as the first and last columns of $M_c(T_3)$ and replacing this matrix by a matrix $M'(T_3)$ with six rows and six columns (see Table 9.7).

9.6.2. *Generalization: Boolean lattices as n-cubes*

In the case of Boolean lattices T_n, the Hasse diagrams are associated with the T_n are n-cubes. So, we may observe that it is possible to directly compute the edges of Boolean lattice incomparability graphs, without using the previous method or the comparability graph Boolean representation (see Parrochia 2020).

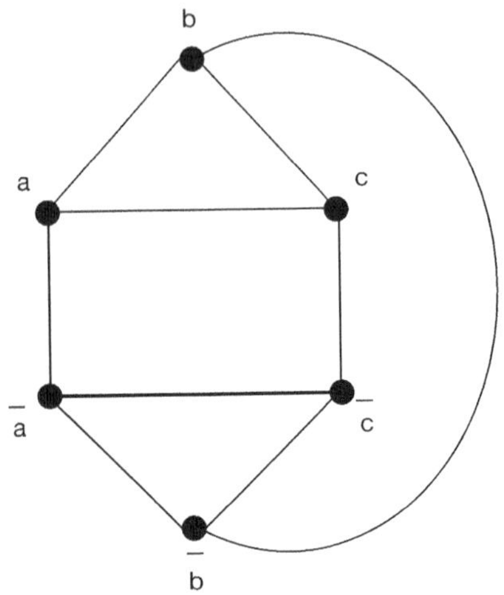

Figure 9.8. *Incomparability graph of T_3*

$M'(T_3) =$

	a	c	$\bar{b}$	b	$\bar{c}$	$\bar{a}$
a	0	1	0	1	0	1
c	1	0	0	1	1	0
$\bar{b}$	0	0	0	1	1	1
b	1	1	1	0	0	0
$\bar{c}$	0	1	1	0	0	1
$\bar{a}$	1	0	1	0	1	0

Table 9.7. *Adjacency matrix of the incomparability graph of T_3*

THEOREM 9.4.– *It is well known that the number $|U|$ of edges of an n-cube with 2^n vertices is given by the formula:*

$$|U| = n \times 2^{n-1}.$$

THEOREM 9.5.– *The number $|U'|$ of edges of the incomparability graph of a Boolean lattice T_n with n vertices is obtained by subtracting from the cardinal $\overline{|U|}$ (number of edges of the complementary graph of T_n), the total sum of the n-faces ($n \geq 2$) of the corresponding n-cube.*

PROOF.– Let T_n be a Boolean lattice with 2^n elements and $\overline{|U|}$ be the number of edges of its complementary graph. To find the number $|U'|$ of edges of the incomparability graph of T_n, we must remove from the graph $G(X, \overline{U})$, complementary to T_n, the ordered pairs of non-adjacent vertices in T_n. These ones are in bijection with the successive n-faces of the n-cube associated with T_n. Let k be the dimension of the faces. Given a fixed n, for each value of k, the number of these faces is $N_{n,k} = 2^{n-k}\binom{n}{k}$ (see Sommerville 1929, p. 29; Coxeter 1973, p. 122). We then have:

$$|U'| = \overline{|U|} - \sum N_{n,k}.$$

□

Some special formulas for computing the number of faces of an n-cube:

– two-sided: $n(n-1)2^{n-3}$;

– three-sides: $n(n-1)\frac{n-2}{3}2^{n-4}$;

– four-faces: $n(n-1)(n-2)\frac{n-3}{3}2^{n-7}$.

In the general case, it suffices to calculate $N_{n,k}$.

Examples

For the lattice T_3 (3-cube), we have, by the above formulas, 6 two-faces and 1 three-face, which gives 6 + 1 = 7. As $\overline{|U|} = 16$, we check that we get $|U'| = 16 - 7 = 9$.

For the lattice T_4 (4-cube), we have, by the same formulas, 24 two-faces, 8 three-faces and 1 four-faces, which gives 24 + 8 + 1 = 33. Knowing that here $\overline{|U|} = 88$, we get $|U'| = 88 - 33 = 55$.

For the lattice T_5 (5-cube), we would obtain: 80 two-faces, 40 three-faces, 10 four-faces and 1 five-face, which gives: 80 + 40 + 10 + 1 = 131. Knowing that here $\overline{|U|} = 416$, we get $|U'| = 416 - 131 = 285$.

Finally, for the lattice T_6 (6-cube), we would obtain: 240 two-faces, 160 three-faces, 60 four-faces and 12 five-faces and 1 six-faces, which gives: 240 + 160 + 60 + 12 + 1 = 473. Knowing that here $\overline{|U|} = 1{,}824$, we get $|U'| = 1824 - 473 = 1{,}351$.

Table 9.8 shows a summary of the characteristics of T_n, the complementary graphs and the corresponding incomparability graphs (appear successively in the matrix, the lattices T_n, the corresponding product Boolean sets B^n, the number of vertices $|X|$ of the associated graphs, the number of edges $|U|$, the number of edges $\overline{U}$ of the complementary graphs, and finally, the number of vertices $|X'|$ and the number of edges U' of the corresponding incomparability graphs.

T_n	B^n	$\|X\|$	$\|U\|$	$\|\overline{U}\|$	$\|X'\|$	$\|U'\|$
T_2	$\{0,1\}^2$	4	4	2	2	1
T_3	$\{0,1\}^3$	8	12	16	6	9
T_4	$\{0,1\}^4$	16	32	88	14	55
T_5	$\{0,1\}^5$	32	80	416	30	285
T_6	$\{0,1\}^6$	64	192	1824	62	1351

Table 9.8. *Characteristics of some lattices, complementarity graphs and incomparability graphs*

Note finally that we can always get the number of edges of a comparability graph in another way. If we call V the number of edges corresponding to non-explicit order relations in a poset P, it is clear that the number of edges $|E|$ of the comparability graph of P is:

$$|E| = |U| + |V|.$$

And we have also, of course (see Theorem 9.3):

$$|U| + |V| = \binom{n}{2} - N.$$

9.7. Consequences for negative theology

We can extend these remarks with more philosophical considerations: the path of knowledge by complementarity or negation of the phenomenal existence has been explored since the most remote antiquity in this current of thought called "negative theology". According to the remark of Philo of Alexandria (20–45 BC), the most famous philosopher of the Judeo-Platonic school, God transcends all knowledge and remains in principle forever unspeakable (*arrêtos*) and incomprehensible (*akatalêptos*). We cannot therefore say what He is, we can only say what He is not. This thesis will be found later in Clement of Alexandria (∼150–∼215) and in Gregory of Nyssa (∼335–∼395), one of the "Cappadocian" fathers, and it will be especially defended by Pseudo-Denys the Areopagite, a Christian author of the 6th century, who was also an heir of Neoplatonism (Plotinus, Porphyry of Tyre, Jamblicus and Damascius) (see Vannier 1998). Thus emerges a way called "apophatic" (from the Greek "apophasis", negation) whose distant origin is undoubtedly Plato's *Parmenides*. As in the case of a fruit whose core is reached by removing the flesh, it aims to describe what we do not know and cannot know by aligning negations of what we do know: firstly, it is not absolutely a question of seeking to *know* Him who transcends all things, it is above all a question of *uniting oneself* with Him; secondly, the use of negation applies not only to the discourse (and its predicative structures) but also to the words themselves, which induces oxymorons or contradictions in the terms which approach poetic formulas. Thus, in Dionysius the Areopagite – who takes up Gregory of Nyssa – the famous "very luminous darkness of a silence full of profound teachings: marvelous darkness which shines in splendid flashes, and which, not being cannot be seen nor grasped, it floods with the beauty of its fires the sanctified minds" (see Denys L'Aéropagite 1846, p. 276). This use of contradiction will be later systematized by N. de Cues; finally, the idea of a god who would be the complementary of the complementary (since he is accessible neither to the senses nor to the understanding) must be moderated. In reality, for pseudo-Denys, there is nothing of what is, but there is also nothing of what is not, and, in this sense, is not only above all our affirmations, but above all our negations too. In any case, "one must make of Him neither affirmation nor absolute negation" (see Denys L'Aéropagite 1846, p. 286). So we can see that Denys oscillates here between the definition of what we would call today a *weak order* and that of what we now characterize as a *pseudo-complement*, without being able to specify much more.

This tradition of thought, which would later lead to the Rhenish mysticism of the 13th and 14th centuries – notably Master Eckhart (1260–1328) and his disciple Tauler – and then to Flemish mysticism (Hadewijch of Antwerp, Ruysbroek), has at the outset, as we can see, a rational background. The present study, however, reveals a certain number of its limits, briefly mentioned in a previous work (see Parrochia 2018) and which we will allow ourselves, to finish, to enumerate.

Let us admit, as Dipert wanted (see Dipert 1997) that the structure of the world can be that of a graph (asymmetric in this case). This thesis, although much discussed as we have seen (see in particular Shackel (2011, p. 11) and Oderberg (2011)), perhaps allows us to appreciate more precisely the approach of negative theology by giving it a well-defined framework. In order to define the complementary of anything, it is necessary that this complementary exists, i.e. that the thought can be associated with a well-complemented space. But the mathematical theory of orders reveals different problems as seen below.

1) Some graphs, said to be self-complemented, have the same structure as their complement, except for the permutation of the labels indexing the vertices of the graph. If the structure of the world were self-complemented in this way, not only would there obviously be no way beyond, but the movement of knowledge would be caught in a closed space. The father of structuralism, Claude Lévi-Strauss, once supported such a thesis with much clarity:

> The Universe having signified long before we begin to know what it meant... it results also that it has signified, from the beginning, the

> totality of what the humanity can expect to know. What is called the progress of the human mind and, in any case, the progress of scientific knowledge, can only consist in rectifying cuttings, in proceeding to regroupings, in defining belongings and in discovering new resources, within a closed totality complementary with itself (see Levi-Strauss 1950, p. 30)[3].

It is obvious that, in such a case, the negative of what exists being none other than the existing itself, what escapes the self-complemented existing can only be the empty set.

2) The strict weak orders have as complements total pre-orders and vice versa. If the structure of the world were a strict weak order (antisymmetric, transitive) with transitivity of incomparability, the complementary of what is – different from what is, in such a case – would thus be a total pre-order (reflexive, transitive). Conversely, if the world could be reduced to a total pre-order, its complementary would be a weak order. Hence, in both cases, the impossibility of ordering certain elements.

3) If the structure of the world were to be a total or linear order (chain), its complementary, as we have seen, would be a null graph or an antichain, and thus a set of absolutely discontinuous elements. We had proposed, in the past, to model it by using interval graphs (Parrochia 2004, pp. 182–183), which are, in fact, the complements of comparability graphs. Since we are talking about discontinuity, we will note that the theory of graphs gives a precise elaborate definition of the notion of the "hole". A hole is an odd cycle without chord with at least five vertices, while an anti-hole is the complementary graph of a hole. Thus, the Pythagorean pentagram is a hole. On the other hand, incomparability graphs are without holes or anti-holes. Indeed, incomparability graphs are perfect graphs, and a graph G is precisely perfect if and only if it contains no holes and no anti-holes as subgraphs. This conjecture, stated by Berge in 1960, became a

3 It will not be objected that only the mythical thought operates in this closed space with combinations comparable to the one of the "bricolage", while the scientific space would be quite other. Certainly, the concept is an operator of opening a little more powerful than the image and which seems to engage an apparently more open space. But, in reality, as Lévi-Strauss explains very well, the role of the concept does not extend nor renew the signification, but "is limited to obtain the group of its transformations" (Levi-Strauss 1962, p. 30). These considerations illuminate the approach of Jacques Derrida, of a new day. The "deconstruction", a kind of negative theology without God, supposes that, by dismantling and unbinding operated within the philosophical language, it would be possible to make another space appear compared to the one of metaphysics. But this hope is doomed to failure if the space of metaphysics, which is in reality identified with that of signification or meaning, is – as Lévi-Strauss thinks – self-complemented.

theorem only 42 years later, in May 2002, thanks to Maria Chudnovsky and Paul Seymour. The proof of this theorem is now 148 pages long.

4) If the graph of the world were to be only a partially ordered set (poset), then we must note that the complementary of an order – even a partial one – does not always exist. Thus, some lattices – for example, distributive lattices – are not complementary.

5) When a lattice is complemented, the complement of an element is not necessarily unique. In the lattice of Figure 9.5A, an element may have two complements. The complement of what it is may therefore not be unique: so it is not *one* entity that would appear in this way, but *several*: a real problem for monotheism...

6) Some partial orders, such as pseudo-complemented or relatively pseudo-complemented lattices, admit only "pseudo-complements" and not complements. In a lattice T with a lower element 0, an element $x \in T$ is said to have a pseudo-complement if there exists a larger element $x^* \in T$, disjoint from x with the property that $x \wedge x^* = 0$. In this case, it only means that $x^* = \text{Max}\ \{y \in T\ |x \wedge y = 0\}$. This, again, gives rise to a class where one element is only privileged among others.

7) Only Boolean lattices (distributive and complemented, as we have seen), give rise to one and only one complement for any element.

8) In the case of matroids or of their associated hypergraphs, the passage to the complement leads either to an autocompletion or to an extremely large number of alternative structures.

9) In the case of simplicial complexes, it is even worse: apart from the cases mentioned at the end of the previous section, the notion of complementarity loses its meaning and the number of alternative structures becomes enormous.

Consequently – under the hypothesis that the whole of what is perceived or understood can be reduced to one of the preceding structures – the approach of negative theology (or of any other form of thought which would claim to define some entity by its complement) could only be valid in this space of thought corresponding to a "well-complemented" partial order, i.e. on order of Boolean type. In any other circumstance, this approach would fail.

10

From Fuzzy Graphs to Neutrosophic Graphs

This short chapter is devoted to the study of some generalizations of graphs that we did not want to completely ignore and which may have some philosophical importance. As Pakistani mathematician Muhammad Akram (2018, p. 1) well summarized, Zadeh introduced the degree of membership for an element of a set, or the degree of truth T associated with this membership in 1965. Symmetrically, for such an element, Atanassov introduced its degree of non-membership or its degree of falshood F in 1983, thus defining the notion of an intuitionistic fuzzy set. In 1995, the Romanian-American mathematician Florentin Smarandache introduced the degree of indeterminacy or neutrality I of a concept or a statement as an independent component, defining by the way what he calls a *neutrosophic set*, based on three components: $\{T, I, F\}$ = (Truth, Indeterminacy, Falsity).

10.1. Fuzzy sets

Let us first quickly recall what a fuzzy subset is in the sense of Zadeh.

Let E be a set and $A \subset E$ be a subset of E, and we usually indicate by the symbol "$\in$" that an element x of E *belongs* to A. We can also use, in an equivalent way, what we call its *membership function*, denoted $\mu_A(x)$, whose value indicates whether or not x belongs to A. We have:

$$\mu_A(x) = 1 \quad \text{if} \quad x \in A.$$

$$\mu_A(x) = 0 \quad \text{if} \quad x \notin A.$$

Suppose, for example, a finite set E with five elements:

$$E = \{x_1, x_2, x_3, x_4, x_5\};$$

and a three-element A subset:

$$A = \{x_1, x_3, x_5\},$$

we can express A as the following set of pairs $((x_i, \mu_A(x_i))$:

$$A = \{(x_1, 1), (x_2, 0), (x_3, 1), (x_4, 0), (x_5, 1\}.$$

In this example, the elements of A belong or not to the set E and the membership function takes its values in $\{0, 1\}$.

In the case of a fuzzy subset, we will admit that the membership function can take its values in $[0, 1]$. We therefore introduce the mathematical concept defined by an expression of the following type:

$$\widetilde{A} = \{(x_1|0,2), (x_2|0), (x_3|0,3), (x_4|1), (x_5|0,8)\}.$$

x_i is an element of the frame E and the number placed after the bar indicates the value of the membership function of this element. The mathematical concept introduced here will be called "fuzzy subset" of E and denoted $\widetilde{A}$.

This notion of fuzzy membership makes it possible to express that an element belongs "more or less" to a set. It could be useful in philosophy in areas such as aesthetics (where we can suggest that something is "more or less" beautiful), or in morals (where we can express that such and such an action is "more or less" correct). In the history of philosophy, we can also imagine expressing such and such an author (respectively, text) has "more or less" influenced such and such another author (respectively, text). There are certainly many other applications (the reader will find some in Parrochia (1993a)).

Even for the most convinced philosophers of the existence of arithmetic rationality, there are limits to the laws of addition (and subtraction) when applied to everyday life. Everyone knows this famous text by Hobbes, premonitory of modern computing, which identifies reason with a calculation:

> When a man Reasoneth, he does nothing els but conceive a summe totall, from Addition of parcels; or conceive a Remainder, from Substraction of one summe from another: which (if it be done by Words,) is conceiving of the consequence of the names of all the parts, to the name of the whole; or from the names of the

> whole and one part, to the name of the other part. And though in some things, (as in numbers,) besides Adding and Substracting, men name other operations, as Multiplying and Dividing; yet they are the same; for Multiplication, is but Addition together of things equall; and Division, but Substracting of one thing, as often as we can. These operations are not incident to Numbers onely, but to all manner of things that can be added together, and taken one out of another. For as Arithmeticians teach to adde and substract in Numbers; so the Geometricians teach the same in Lines, Figures (solid and superficiall,) Angles, Proportions, Times, degrees of Swiftnesse, Force, Power and the like; The Logicians teach the same in Consequences Of Words; adding together Two Names, to make an Affirmation; and Two Affirmations, to make a syllogisme; and Many syllogismes to make a Demonstration; and from the Summe, or Conclusion of a syllogisme, they substract one Proposition, to finde the other. Writers of Politiques, adde together Pactions, to find mens Duties; and Lawyers, Lawes and Facts, to find what is Right and Wrong in the actions of private men. In summe, in what matter soever there is place for Addition and Substraction, there also is place for Reason; and where these have no place, there Reason has nothing at all to do (see Hobbes 1651, Chapter V).

Faced with the need to draw up, for a man, a balance sheet of good and evil, of his good and bad actions, Hobbes, however, hesitates: he is not sure that in legal and moral matters, for example, a strict application of arithmetic is still relevant: Does a "righteous" man lose this qualifier if he has ever done one unjust action? Conversely, will a deeply unrighteous man lose this character if, for once, and under the influence of fear, he refrains from committing one more unjust action? We feel that fuzziness slips under the apparent rigor of number theory, and Hobbes would have needed, no doubt, to give a more precise formulation, the fuzzy sets theory.

We would say the same for Malebranche, for whom "God compares and weighs everything". Acting according to the most general laws, it sometimes creates local imbalances where nature is not compensated by grace. Thus, a sinner who has four degrees of lust may receive only three degrees of "spiritual delight". Some will be granted more grace, others less, for reasons that only God knows and that are aimed at the general good. We must also distinguish between the grace of the creator and the grace of Jesus Christ. The latter does not cease to act, even when concupiscence is not reduced. Moreover, it combines in a complex way with charity. According to Malebranche, there are also too many parameters, for us, to be able to make, here again, a precise assessment:

> The different degrees of light, of charity, of concupiscence, the different degrees of freedom combining at all times, in an infinity of ways with the different degrees of actual pleasures, and these pleasures having their effect only according to the ratio they have with the dispositions of mind and heart: it is evident that no finite mind can judge with any assurance of the effect that a particular grace must produce in us. For besides the fact that the combination of all that contributes to make it effective or to produce its effect contains something infinite, this combination is not similar to that of springs and moving forces whose effects are always infallible and necessary. Thus, no spirit can discover what is going on in the heart of man (see Makinson 1958, section 37, p. 265).

Blurry insinuates itself, here again, into the mechanics of grace and the balance sheet of goods and pains.

In general, philosophers continue to evolve in the dark, or if not in the dark, at less in the "fuzzy". But now they could do it formally.

10.2. Fuzzy graphs

As we know, a *graph* $G(V, E)$ is an ordered pair comprising a set V of vertices or nodes or points together with a set E of edges or arcs or lines, which are two-element subsets of V. A priori, we therefore do not see where fuzziness could be lodged. But we can start from another definition, namely, the idea of a set of pairs defined over a product set, which is the Berge's definition.

DEFINITION 10.1.– *A graph in the sense of Berge is such that:*

$$E_1 = E_2 = E \quad \text{(countable)};$$

and it is formed by the subset of couples:

$$(x, y) \in G \subset E \times E;$$

such as:

$$G \cap \bar{G} = \emptyset;$$

and:

$$G \cup \bar{G} = E \times E.$$

For such graphs, which are obviously only a particular case of the usual notion of graph, we can define a generalization to fuzzy graphs.

DEFINITION 10.2.– *Consider two sets E_1 and E_2 and denote by x an element of E_1 and by y an element E_2. The set of pairs (x, y) defines the product set $E_1 \times E_2$. The fuzzy subset such as:*

$$\forall(x, y) \in E_1 \times E_2 : \mu_{E_1 \times E_2}(x, y) \in M;$$

where M is the membership set of $E_1 \times E_2$, is called a fuzzy graph.

Example: The following matrix (see Table 10.1) and graph (see Figure 10.1) are usual Berge's ones.

Γ	A	B	C
A	0	1	0
B	1	0	0
C	0	1	1

Table 10.1. *Incidence matrix of the Berge's graph G*

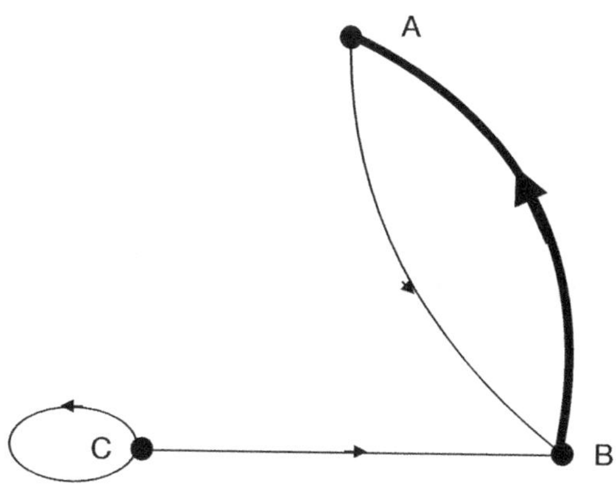

Figure 10.1. *Berge's graph*

We can now define a fuzzy generalization of them with Table 10.2 and Figure 10.2 (see Kaufmann and Boulaye 1978, pp. 41–46):

Γ	A	B	C
A	0,5	1	0
B	0	0	0,5
C	1	1	0

Table 10.2. *Incidence matrix of the fuzzy graph* G'

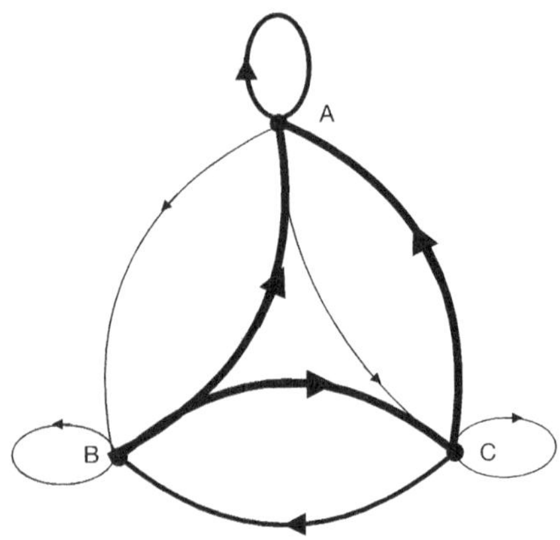

Figure 10.2. *Fuzzy graph* G'

10.3. Intuitionistic fuzzy set theory

As we have seen, fuzzy set theory allows the gradual assessment of the membership of elements in a set. This is described with the aid of a membership function valued in the real unit interval $[0, 1]$. During the 1980s, Bulgarian mathematician and computer scientist Krassimir Todorov Atanassov (b. 1954) developed a generalization of Zadeh's fuzzy set theory.

The theory of intuitionistic fuzzy sets further extends the previous concepts by allowing the assessment of the elements by two functions: μ for membership and ν for non-membership. Both of them belong to the real unit interval $[0, 1]$, and their sum $\mu \pm \nu$, of course, belongs to the same interval.

Intuitionistic fuzzy sets generalize fuzzy sets, since the indicator functions of fuzzy sets are special cases of the membership and non-membership functions of fuzzy sets. In the case when the strict equality exists, say:

$$\nu = 1 - \mu$$

the non-membership function fully complements the membership function to 1, not leaving room for any uncertainty.

Let us have, as in fuzzy set theory, a fixed universe E and let A be a subset of E. Let us construct the set:

$$A^* = \{\langle x, \mu_A(x), \nu_A(x)\rangle | x \in E\},$$

where $0 \leq \mu_A(x) + \nu_A(x) \leq 1$.

We will call the set A^* intuitionistic fuzzy set (IFS).

In the publications on IFS, authors mainly deal with the concept of intuitionistic fuzzy set A^* rather than with fixed set A. Hence, for the sake of simplicity, major publications presenting the very definition of the concept often use notation A instead of A^* (see Atanassov 1999; Atanassov 1986). There exists another definition of an intuitionistic fuzzy set, mathematically more precise than the previous one. It is the following:

$$A^* = \{\langle x, \mu_A(x), \nu_A(x)\rangle | x \in E \& 0 \leq \mu_A(x) + \nu_A(x) \leq 1\}.$$

but it is also a more complex one and is never used, as of 2008.

In this theory, in addition to the functions $\mu : E \to [0,1]$ and $\nu : E \to [0,1]$, we also define: $\pi : E \to [0,1]$, through $\pi(x) = 1 - \mu(x) - \nu(x)$ corresponding to the degree of uncertainty (indeterminacy, etc.). Of course, for every ordinary fuzzy set A, $\pi_A(x) = 0$, for each $x \in E$.

For every two intuitionistic fuzzy sets A and B, various relations and operations may be defined, the most important of which are:

– inclusion: $A \subset B$ iff $(\forall x \in E)(\mu_A(x) \leq \mu_B(x)\ \&\ \nu_A(x) \geq \nu_B(x))$, $A \supset B$ iff $B \subset A$;

– equality: $A = B$ iff $\forall x \in E)(\mu_A(x) = \mu_B(x)\ \&\ \nu_A(x) = \nu_B(x))$;

– classical negation: $\bar{A} = \{\langle x, \nu_A(x), \mu_A(x)\rangle | x \in E\}$;

– conjunction: $A \cap B = \{\langle\ \min((\mu_A(x), \mu_B(x)),\ \max((\nu_A(x), \nu_B(x))\rangle | x \in E\}$;

– disjunction: $A \cup B = \{\langle\ \max(\mu_A(x), \mu_B(x)),\ \min((\nu_A(x), \nu_B(x))\rangle | x \in E\}$.

These operations and relations are defined similarly to these from the fuzzy set theory. Probably more interesting are the modal operators that can be defined over intuitionistic fuzzy sets. These have no analogue in fuzzy set theory.

10.4. Neutrosophy

As Akram remarked, “Fuzzy set theory and intuitionistic fuzzy set theory are useful models for dealing with uncertainty and incomplete information. But they may not be sufficient in the modelling of indeterminate and inconsistent information encourtered in real world” (see Akram 2018, p. 1). Hence, Smarandache invented *Neutrosophy*. According to him, neutrosophy is a new branch of philosophy “which studies the origin, nature, and scope of neutralities, as well as their interactions with different ideational spectra” (Smarandache 2002, p. 297).

We may precise these statements by the following:

Let us note by <A> an idea, or proposition, theory, event, concept, entity. Then:

1) <Non-A> means what is not <A>;

2) <Anti-A> is the opposite of <A>;

3) <Neut-A> means what is neither <A> nor <Anti-A>, i.e. neutrality in between the two extremes;

4) <A’> is a version of <A>.

Remark

<Non-A> is different from <Anti-A>. For example:

1) if <A> = white, then <Anti-A> = black (antonym), but <Non-A> = green, red, blue, yellow, black, etc. (any color, except white);

2) <Neut-A> = green, red, blue, yellow, etc. (any color, except white and black);

3) <A’> = dark white, etc. (any shade of white).

Relations:

1) <Neut-A> ∪ <Neut-(Anti-A)>, i.e. neutralities of <A> are identical with neutralities of <Anti-A>;

2) <Non-A> ⊃ <Anti-A>;

3) <Non-A> ⊃ <Neut-A;

4) <A> ∩ <Anti-A> = ∅;

5) <A> ∩ <Non-A> = ∅.

<A>, <Neut-A>, and <Anti-A> are disjoint two by two. <Non-A> is the completeness of <A> with respect to the universal set.

But since, in many cases, the borders between notions are vague or imprecise, it is possible that <A>, <Neut-A>, <Anti-A> (and <Non-A> of course) have common parts two by two.

In this context, the main principle (MP) states that:

(MP) Between an idea <A> and its opposite <Anti-A>, there is a continuum-power spectrum of neutralities <Neut-A>.

The three functions T, I and F by which neutrosophic sets are identified take their values in real standard or non-standard subset of unit interval $]^{-}0, 1^{+}[$ where $^{-}0 = 0 - \epsilon, 1^{+} = 1 + \epsilon, \epsilon$ is an infinitesimal number.

So we get the fundamental thesis (Th):

(Th) Any idea <A> is $T\%$ true, $I\%$ indeterminate and $F\%$ false, where $T, I, F \subset]^{-}0, 1^{+}[$.

This extent logic beyond existing generalizations such as many-valued logic or even the quantum logic from C. F. von Weizsäcker.

Recall that Jan Łukasiewicz developed a system of three-valued logic in 1920. He generalized the system to many-valued logics in 1922 and went on to develop logics with $\aleph_0$ (infinite within a range) truth values. Kurt Gödel developed a deductive system, applicable for both finite- and infinite-valued first-order logic.

On the other hand, C. F. von Weizsäcker developed a kind of quantum logic named "logic of complementarity", in order to find the unity of physics (see Weizsäcker 1971; Parrochia 2021). According to Max Jammer, this logic is constructed as a modification of the logic of contingent propositions, in particular that of simple alternatives (*einfache Alternative*), as it is presented in experiments such as that of Young's slits, when the particle passes either through slit 1 or through slit 2. Von Weizsäcker derives the rules of his logic from the very situation which is described by quantum mechanics.

As we know, the state function ψ, behind the diaphragm formed by the slits, is only determined by the two complex numbers u and v defined by the equation:

$$\psi = u\phi_1 + v\phi_2$$

where $uu^* + vv^* = 1$, ϕ_1 and ϕ_2 propositions characterized by (u, v) and which differ from a_1 and of a_2 are said to be "complementary to a_1 and a_2".

The complementarity therefore takes on a purely logical character here: if one of the two propositions is true or false, the complementary is neither true nor false.

We find an illustration of this idea in the quantum theory of spin-$\frac{1}{2}$ particles, which seems to have inspired it. In this example, the vector (u, v) is simply the two-component Pauli spinor. If a given magnetic field is directed in the positive direction z, then the original alternative lies between the two propositions:

a_1: the spin is in the positive direction z.

a_2: the spin is in the negative direction $-z$.

To any spinor, (u, v) then corresponds a direction of spin orientation, which, if a_1 has the value u and a_2 has the value v, can be defined, in terms of direction cosines, by the formulas:

$$x = uv^* vu^*; \qquad y = i(uv^* vu^*);$$

$$z = uu^* - vv^*;$$

$$x^2 + y^2 + z^2 = 1,$$

and, in terms of polar angle θ and azimuthal angle ϕ, by the equations:

$$u = \cos\frac{\theta}{2}\exp(-\frac{i\phi}{2}), \qquad v = \sin\frac{\theta}{2}\exp(-\frac{i\phi}{2}).$$

If θ is neither 0 nor π, then the proposition "The spin has the direction (θ, ϕ)" is complementary to the propositions a_1 and a_2.

In the wake of this complementarity logic of von Weizsäcker, it should be possible to further generalize Smarandache's approach.

The concept of expressing truth values as real numbers in the range between 0 and 1 can bring to mind the possibility of using complex numbers to express truth values. These truth values would be, as in von Weizsäcker's logic, couples of numbers, to be taken between $(0, 0)$ and $(1, 1)$.

In this new version, the three functions T, I and F by which neutrosophic sets are identified take no more their values in real standard or non-standard subset of unit interval]–0, 1+[but in complex subset of $[z, w]$ with $z = (0, 0)$ and $w = (1, 1)$, which is not exactly an "interval" but a square region of the complex plane.

So we get the fundamental generalized thesis (GTh):

(GTh) Any idea <A> is $T\%$ true, $I\%$ indeterminate and F% false, where $T, I, F \subseteq [z, w]$.

Obviously, the philosophical problem is less able to measure the degree of truth, indeterminacy or falsity of an idea than to know how an idea can be true, false or indeterminate. This supposes, each time, an argument and hence, as far as we know, philosophy cannot be entirely mathematized.

Furthermore, it is clear that Smarandache's theory, assuming that any statement can be true, false and indeterminate (at least in a certain percentage), takes skepticism to its extremes. Moreover, it empties of its meaning of philosophical reflection, which supposes that we assume clear choices and that we defend them without imagining, even before they are contested, that they are in any case partly false.

Anyway, Smarandache has introduced neutrosophic graphs (Vasantha Kandasamy et al. 2015; Smarandache 2020) and Akram (2018) has studied some of them, the "single-valued neutrosophic graphs". Although highly controversial, Smarandache's ideas, or at least some of them, were eventually accepted and published by a major publisher. Finally, let us say a few words about this last theory which deals with "single-valued neutrosophic graphs".

10.5. Single-valued neutrosophic sets and graphs

Let us define formally first a "single-valued neutrosophic set".

DEFINITION 10.3.– *Let X be a space of points (objects). A single-valued neutrosophic set A on a nonempty set X is characterized by a truth-membership function $T_A{:}X \rightarrow [0,1]$, indeterminacy-membership function $I_A{:}X \rightarrow [0,1]$ and a falsity-membership function $F_A{:}X \rightarrow [0,1]$. Thus, $A = \{\langle x, T_A(x), I_A(x), F_A(x)\rangle | x \in X\}$. There is no restriction on the sum of $T_A(x), I_A(x), F_A(x)$ for all $x \in X$.*

When X is discrete, a single-valued neutrosophic set A can be written as:

$$A = \sum_{i=1}^{n} \langle (T(x_i), I(x_i), F(x_i))/x_i, x_i \in X\rangle.$$

Examples: Consider a universe of discourse $X = \{x_1, x_2, x_3\}$, where x_1, x_2, x_3 describes some properties of objects.

We can easily imagine that this kind of formalism will be useful in marketing, especially if it is a question of describing the capability, trustworthiness or prices of the objects.

But it is not forbidden to use it for something else. For example, we can assume that x_1 designates the explanatory power of a statement (respectively, of a set of statements), x_2 designates its (respectively, their) reliability and x_3 designates its (respectively, their) interest for thought (it is assumed that experts agreed, at some point, to set these values in a consensual manner).

Or again, we may want to assess the citizen's representation of the three French republican values of liberty, equality and fraternity, to which objective sociological surveys would make it possible to attribute degrees of relevance, indifference or rejection. We can then reason more formally on these different elements.

Each time, the set A will be formed of quadruplets of the type:

$$A = \{\langle x_1, a, b, c\rangle, \langle x_2, c, d, e\rangle, \langle x_3, f, g, h\rangle\};$$

where $a, b, c, d, e, f, g, h \in [0, 1]$.

Let us now define a "single-valued neutrosophic relation".

DEFINITION 10.4.– *A single neutrosophic relation on a non-empty set X is a single-valued neutrosophic subset of $X \times X$ of the form:*

$$B = \{(yz, T_B(yz), I_B(yz), F_B(yz)) : yz \in X \times X\};$$

where $T_B : X \times X \to [0, 1], I_B : X \times X \to [0, 1], F_B : X \times X \to [0, 1]$ denotes the truth-membership function, indeterminacy-membership and falsity-membership function of B, respectively.

We define also the complement B^c of B, the inverse B^{-1} of B and the conditions for a relation to be reflexive, symmetric or anti-symmetric and transitive (for details, see Akram (2018, pp. 5–6)).

DEFINITION 10.5.– *A single-valued neutrosophic graph on a non-empty set X is a pair $G = (A, B)$, where A is the single-valued neutrosophic set in X and B is the single-valued neutrosophic relation on X such that:*

$$T_B(xy) \leq \min[T_A(x), T_A(y)];$$

$$I_B(xy) \leq \min[I_A(x), I_A(y)];$$

$$F_B(xy) \leq \max[F_A(x), F_A(y)];$$

for all $x, y \in X$.

We borrow the following example from Akram (2018):

Let $G^*(X, E)$ be a crisp graph such that $X = \{a, b, c, d, e, f\}$ and $E = \{ab, ac, bd, cd, be, cf, ef, bc\}$. Let A and B be the neutrosophic sets of X and E, respectively, as shown in Table 10.3.

A	a	b	c	d	e	f		
T_A	0.2	0.3	0.4	0.3	0.5	0.4		
I_A	0.5	0.4	0.5	0.6	0.5	0.6		
F_A	0.7	0.6	0.4	0.8	0.6	0.6		
B	ab	ac	bd	cd	be	cf	ef	bc
T_B	0.2	0.1	0.2	0.3	0.2	0.1	0.4	0.2
I_B	0.4	0.4	0.2	0.2	0.3	0.4	0.4	0.3
F_B	0.7	0.5	0.6	0.7	0.5	0.5	0.5	0.6

Table 10.3. *Neutrosophic sets*

By simple calculation, it is easy to see that $G = (A, B)$ is a neutrosophic graph as shown in Figure 10.3.

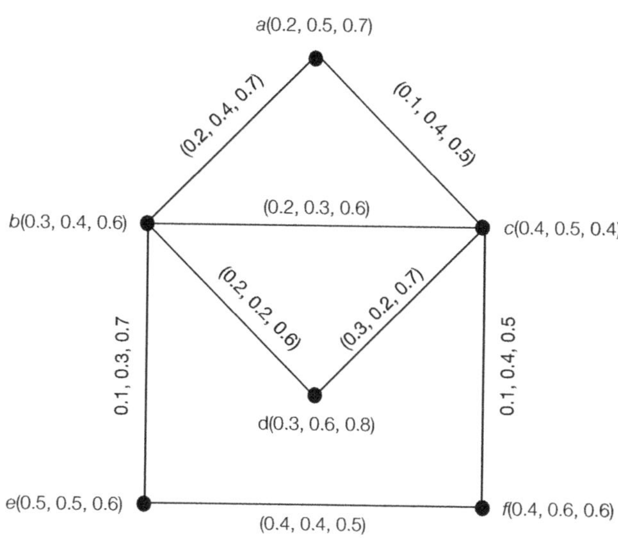

Figure 10.3. *Neutrosophic graph*

Many concepts of ordinary graph theory (direct graphs or digraphs, complete graphs, planar graphs, regular graphs, order, size, degree of a graph, etc.) generalize to neutrosophic graph theory.

This extension of graph theory can be useful in social sciences.

Conclusion

Graphs and Polyhedra of Ideas

The idea of putting thought into graphs is highly justified, although we do not think, like Dipert, that the structure of the world itself is a kind of graph.

Indeed, it is not a new idea, although the graphs in question sometimes had more concrete forms – for example, those of polyhedra – and were not often taken seriously.

For example, Alfred Jarry's affection for polyhedra is well known. Several texts are devoted to them by the author, including one – more philosophical than literary – which contains this recommendation: "Let us therefore weigh words, polyhedra of ideas, with scruples like diamonds" (see Jarry 2007).

The formula inspired the French philosopher and epistemologist Gaston Bachelard, who retained it to characterize the scientific spirit. According to the latter, "the scientific mind builds coherent sets of ideas, or, to use Alfred Jarry's fine expression, 'polyhedra of ideas' (see Bachelard 1949, p. 214). But Bachelard does not go further to develop this remarkable intuition. This evocation remains a metaphor for him. Here, we have intended to go beyond.

We must first ask ourselves what a polyhedron is and also how we can classify these objects.

The answer to the first question may seem simple. A polyhedron is informally defined as a three-dimensional geometric shape (a geometric solid) having polygonal plane faces that meet in line segments called edges.

As strange as it may sound, since the concept of polyhedron does not refer to the dimension of the space in which it is located, there is no universally agreed definition of what makes "something" a polyhedron (the heart of the problem comes from the fact that the intuitive notion of polyhedron is not exactly the same depending on whether we have a surface or a volume in the idea).

In order to overcome this difficulty, we introduce the notion of simplex. It can be considered as equivalent to that of a polyhedron in dimension 3, which allows generalizations to higher dimensions. A polyhedron P of dimension p is then the union of a finite set of simplexes S_i of dimension $q_i \leq p$ such that each of the d-faces ($d \leq q_i$) of S_i is an element of P, and such that, for any pair of simplex (S_i, S_j), the intersection $S_i \cap S_j$ is either empty or a $(d-1)$ side common to S_i and S_j.

Thus, a simplex represents a generalizable definition of the intuitive notion of polyhedron. It is the union of its d-faces, and the intersection of any two d-faces of a simplex is either empty or a face of dimension $d-1$. For example, a triangle, which is a 2-simplex, is the union of segments, and the intersection of two adjacent segments is a point that is a vertex of the triangle.

Thus, a polyhedron seems to be constructed from different kinds of elements or entities, presenting a different number of dimensions:

– 3 dimensions: the body is limited by the faces and usually corresponds to the volume included inside;

– 2 dimensions: a face is bounded by a circuit of edges and is usually a planar region called a polygon. The faces put together form the blocky surface;

– 1 dimension: an edge joins one vertex to another and one face to another and is usually a line of some sort. The edges put together form the polyhedral skeleton;

– 0 dimension: a vertex is a corner point;

– -1 dimension: nullity is a kind of non-entity required by abstract theories.

These types of graphs are quite simple in appearance. In reality, the polyhedra actually constitute a real jungle where it is very difficult to tidy up. Among their main characteristics, we can retain the following:

– the number of their faces, which is generally the element that gave them, according to the Greek language, their name: tetrahedron (four faces), pentahedron (five faces), hexahedron (six faces), octahedron (eight faces), dodecahedron (12 faces), icosahedron (20 faces), etc.;

– their convexity or not convexity: a polyhedron is said to be *convex* if any point of any segment joining any two points of the polyhedron belongs to the polyhedron (i.e. if all its diagonals are inside itself). All convex polyhedra satisfy the Euler relation. If f is the number of faces of the polyhedron, a is the number of edges and s is the number of vertices, then the Euler characteristic χ satisfies the relation[1]:

$$\chi \equiv f - a + s = 2$$

– the symmetries that characterize them.

Studying them involves appealing to the concepts of uniformity and regularity, and characterizing them more precisely by means of the group structure.

We will say that a polyhedron has a *uniform vertex* (respectively edge, face) if its vertices (respectively edges, faces) are the same (in the sense that they can be applied to each other).

Now consider the properties of regularity, quasi-regularity and semi-regularity.

A polyhedron is said to be *regular* if it has a uniform vertex, a uniform edge and a uniform face (the uniformity of the vertices and the uniformity of the combined edges imply that the faces are regular). Note that the dual of a regular polyhedron is also regular. A typical example is that of the five Platonic solids (or regular convex polyhedra): tetrahedron, cube, octahedron, regular dodecahedron and regular icosahedron. Plato saw in it the fundamental structure of the "elements" and of the world: the tetrahedron is supposed to be the structure of fire, the cube that of the Earth, the octahedron that of air and the icosahedron that of water. As for the dodecahedron, it is believed to represent the structure of the world. Another example is that of regular Kepler–Poinsot star polyhedra.

1 The Hungarian philosopher Imre Lakatos showed that various discoveries that occurred during the history of mathematics concerning the development of the theory of polyhedra led mathematicians to restrict the validity of this formula to convex polyhedra. He termed the polyhedral counterexamples to Euler's formula *monsters* and distinguished three ways of handling these objects: Firstly, *monster-barring*, by which means the theorem in question could not be applied to such objects. Secondly, *monster-adjustment*, whereby by making a reappraisal of the monster, it could be made to obey the proposed theorem. Thirdly, *exception handling*, a further distinct process. Lakatos tried to establish that no theorem of informal mathematics is final or perfect. This means that we should not think that a theorem is ultimately true; it is just that no counterexample has yet been found. Once a counterexample is found, we adjust the theorem, possibly extending the domain of its validity. This is a continuous way our knowledge accumulates through the logic and process of proofs and refutations (see Lakatos 1976).

Obviously, we can doubt that regularity is a very useful property for the structure of philosophical systems. But this property can weaken. A polyhedron will be said to be *quasi-regular* if it has a uniform edge, but not either with a uniform face or with a uniform vertex.

It will be said to be *semi-regular* if it has a uniform vertex but no uniform face, with each face being a regular polygon. An example of semi-regular polyhedra is that of prisms or antiprisms, and another is that of Archimedean solids (truncations of Platonic solids), whose duals are Catalan polyhedra.

Finally, there are transformation operations that can be performed on the vertices, edges or faces of a polyhedron: stellation, faceting and various truncations.

Polyhedra can still tolerate different generalizations: they can have curved faces or edges, be defined in the complex plane instead of the real plane and may not necessarily be bounded.

Certain combinations of the preceding characteristics may exist: for example, the Johnson solids (92 in number) are convex, non-uniform polyhedra with regular faces.

A polyhedron, the tetrahedron, has the property of being the smallest: it has only four faces, four vertices and six edges.

In his book, *Mysterium cosmographicum*, Kepler developed the theory of regular polyhedra into a model of the universe. He noted that we could intercalate between the orbs of the six planets known at the time (from Mercury to Saturn) as the five Platonic solids (five intervals for six planets).

These polyhedra also explained, by their arrangement, the proportions of the planetary orbs (the relative distances of the planets from the Sun): each solid was inscribed in the orb of a planet and circumscribed to the orb of the planet immediately below. The interlocking was made as follows: the cube between the orbs of Saturn and Jupiter, the tetrahedron between that of Jupiter and Mars, then the dodecahedron between Mars and the Earth, followed by the icosahedron encompassing the orb of Venus, itself circumscribed to the octahedron, which finally surrounded the orb of Mercury.

Kepler thus believed he possessed the structure of the world.

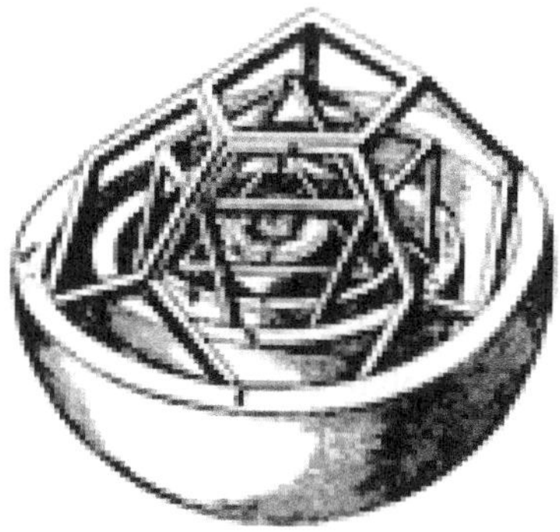

Figure C.1. *The Keplerian structure of the world (source: Wikipedia article: Kepleer)*

Among the other characteristics of the tetrahedron, there is this improbable property that few solids have a figure made up of plane faces, each of which touches the others by an edge. For a long time, it was believed that the tetrahedron was the only one to possess it, but in 1977, the Hungarian mathematician Lajos Szilassi exhibited a strange holed shape made up of 14 vertices, 26 edges and seven faces such that each one was next to the other six.

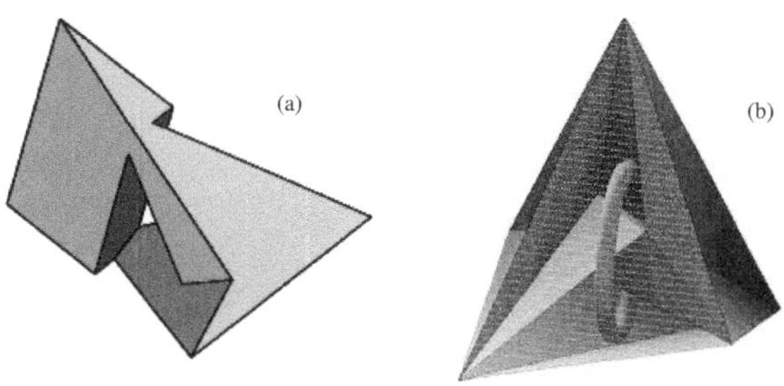

Figure C.2. *Szilassi polyhedron and Császár polyhedron (source: Left: Wikipedia article: Szilassi polyedron; Right: Mathcurve: https://mathcurve.com/polyedres/csaszar/csaszar.shtml). For a color version of this figure, see: http://www.iste.co.uk/parrochia/graphs.zip*

This polyhedron turned out to be the dual of the polyhedron discovered in 1949 by another Hungarian mathematician, topologist Ákos Császár.

The set of vertices and edges of the Császár polyhedron forms a complete graph (denoted as K_7). More generally, if a polyhedron having s vertices, has a edges and f faces corresponds to a surface with t "holes" (i.e. if its Euler characteristic $s - a + f$ is equal to $2 - 2t$), and if its vertices and edges form a complete graph, it has $a = s(s-1)/2$ edges, each face is a triangle and therefore $a = 3f/2$, and we finally obtain $t = \frac{(s-3)(s-4)}{12}$. With t being integer, we must have s congruent to 0, 3, 4, or 7 modulo 12.

This equation holds for the tetrahedron (with $t = 0$ and $s = 4$) and the Császár polyhedron (with $t = 1$ and $s = 7$). The following solution, $t = 6$ and $s = 12$, would correspond to a polyhedron with 44 faces and 66 edges, but such a polyhedron does not exist, and in fact no polyhedron with greater values of t is known.

If a philosopher were to build the structure of a philosophical system – a polyhedron of ideas, as Jarry said – he would therefore have to satisfy these geometric constraints.

In general, graphic organizations constitute essential skeletons for the shaping of our thought and their rigidity imposes restrictions on our combinatorial imagination. It is not bad, in our opinion, that a philosopher should be aware of this at least once, before embarking on the creation of one of those grandiose architectures in which, in the past, we thought that the universe was summed up.

References

Aczel, P. (1988). *Non Well Founded Sets*. Center for study of Language and information, Stanford.

Ageev, A.A. (1996). A triangle-free circle graph with chromatic number 5. *Discrete Math.*, 152, 295–298.

Aharoni, R. and Korman, V. (1992). Greene-Kleitman's theorem for infinite posets. *Order*, 9(3), 245–253.

Akram, M. (2018). *Single-Valued Neutrosophic Graphs*. Springer Nature, Singapore.

Alon, N., Pach, J., Solymosi, J. (2001). Ramsey-type theorems with forbidden subgraphs. *Combinatorica*, 21(2), 155–170.

Apostel, L. (1967). Logique et dialectique. In *Logique et connaissance scientifique*, Piaget, J. (ed.). Gallimard, Paris.

Appel, K. and Haken, W. (1977a). Every planar map is four colorable. Part I. Discharging. *Illinois J. Math.*, 21, 429.

Appel, K. and Haken, W. (1977b). Every planar map is four colorable. Part II. Reducibility. *Illinois J. Math.*, 21, 491.

Apter, A., Di Prisco, C., Henle, J., Zwicker, W. (1989). Filter spaces. II: Limit ultraproducts and iterated embeddings. *Acta Cient. Venez.*, 40 (5), 311–318.

Arruda, A., Chuaqui, A., Da Costa, N. (1980). *Mathematical Logic in Latin America*. North Holland, Amsterdam.

Ascher, M. and Ascher, R. (1997). *Mathematics of the Incas: Code of the Quipu*. Dover Publications, New York.

Asenjo, F.G. (1965). Dialectic logic. *Logique et Analyse*, 8(32), 320–325.

Asenjo, F.G. (1966). A calculus of antinomies. *Notre Dame J. Formal Logic*, 7(1), 103–105.

Asenjo, F.G. and Tamburino, J. (1975). Logic of antinomies. *Notre Dame J. Form. Log.*, 16(1), 17–44.

Atanassov, K.T. (1986). Intuitionistic fuzzy sets. *Fuzzy Sets and Systems*, 20, 87–96.

Atanassov, K.T. (1999). *Intuitionistic Fuzzy Sets*. Springer Physica-Verlag, Heidelberg.

Bachelard, G. (1949). *Le rationalisme appliqué*. P.U.F, Paris.

Bacry, H. (1962). Thomas's classical theory of spin. *Il Nuovo Cimento* (1955–1965), 26(6), 1164–1172.

Bajorath, J. (2004). *Chemoinformatics: Concepts, Methods, and Tools for Drug Discovery, Methods in Molecular Biology*. Human Press, Totowa, NJ.

Balaban, A.T. and Balaban, T.S. (1991). New vertex invariants and topological indices of chemical graphs based on information on distances. *J. Math. Chem.*, 8, 383–397.

Bartozsynski, R. (1971). A note on subjective classifications. *Review of the International Statistical Institute*, 39(I), 39–45.

Bellotti, L. (2005). Woodin on the continuum problem: An overview and some objections. *Logic & Philosophy of Science*, 3(1), 1–12.

Berge, C. (1970). *Graphes et Hypergraphes*. Dunod, Paris.

Berge, C. and Chvátal, V. (1984). *Topics on Perfect Graphs*. North-Holland Pub. Co., Amsterdam.

Beth, E.W. (1948). Les relations de la dialectique à la logique. *Dialectica*, 2(2), 29–39.

Beyssade, J.-M. (1979). *La Philosophie première de Descartes, Le temps et la cohérence de la métaphysique*. Flammarion, Paris.

Biggs, N. (1993). *Algebraic Graph Theory*, 2nd edition. Cambridge University Press, Cambridge.

Biggs, N., Lloyd, E., Wilson, R. (1986). *Graph Theory*, 1736–1936. Oxford University Press, Oxford.

Bird, A. (2007) *Nature's Metaphysics: Laws and Properties*. Oxford University Press, Oxford.

Birkhoff, G. (1940). *Lattice Theory*. American Mathematical Society, Providence, RI.

Blyth, T.S. (2005). *Lattices and Ordered Structure*. Springer Verlag, London.

Bollobás, B. (1990). Almost every graph has reconstruction number three. *J. Graph Theory*, 1–4, 14.

Bollobás, B. and Rival, I. (1979). The maximal size of the covering graph of a lattice. *Algebra Universalis*, 9, 371–373.

Bonchev, D. and Mekenyan, O. (1983). Comparability graphs and electronic spectra of condensed benzenoid hydrocarbons. *Chem. Phys. Lett.*, 98, 134–138.

Bonchev, D. and Mekenyan, O. (1984). Comparability graphs and molecular properties: A novel approach to the ordering of isomers. *J. Chem. Soc. Faraday Trans.*, 2, 695–712.

Bonchev, D., Kamenska, V., Mekenyan, O. (1990a). Comparability graphs and molecular properties III, C9 and C10 alkanes. *Int. J. Quantum Chem.*, 37(2), 135–153.

Bonchev, D., Kamenska, V., Mekenyan, O. (1990b). Comparability graphs and molecular properties IV, generalization and applications. *J. Math. Chem.*, 5, 43–72.

Bondy, J.A. and Murty, U.S.R. (2007). *Graph Theory*. Springer.

Booth, D. and Ziegler, R. (1996). *Finsler Set Theory: Platonism and Circularity, Translation of Paul Finsler's Papers on Set Theory with Introductory Comments*. Birkhäuser Verlag, Basel, Boston, Berlin.

Bourbaki, N. (1971). *Topologie Générale*. Hermann, Paris.

Bourgeois, B. (1970). "Présentation" de Hegel. In *Encyclopédie des sciences philosophiques en abrégé*, Volume 1. Science de la logique, Vrin, Paris.

Bowler, N., Erde, J., Heinig, P., Lehner, F., Pitz, M. (2017). A counterexample to the reconstruction conjecture for locally finite trees. *Bull. London Math. Soc.*, 49(4), 630–648.

Bréhier, E. (1919). L'idée de néant et le problème de l'origine radicale dans le néoplatonisme grec. *Rev. Metaphys. Morale*, 26(4), 443–475.

Bresar, B., Chhangat, M., Klavzar, S., Kovse, M., Mathews, J., Mathews, A. (2008). Cover-incomparability graphs of posets. *Order*, 25, 335–347.

Brink, D. (1989). *Moral Realism and the Foundations of Ethics*. Cambridge University Press, New York.

Caspard, N., Leclerc, B., Monjardet, B. (2007). *Ensembles ordonnés finis: concepts, résultats et usages*. Springer Verlag, Berlin, Heidelberg, New York.

Castellana, M. (2019). *Federigo Enriques e la Nuova Espistemologia*. Pensa Multimedia Editore s.r.l., Lecce.

Cauchy, A.L. (1813). Recherche sur les polyèdres - premier mémoire. *J. Ec. Polytech. Math.*, 9(Book 16), 66–86.

Cayley, A. (1857). On the theory of the analytical forms called trees. *Philos. Mag., Ser. IV*, 13(85), 172–176.

Cayley, A. (1875). Ueber die Analytischen Figuren, welche in der Mathematik Bäume genannt werden und ihre Anwendung auf die Theorie chemischer Verbindungen. *Ber. Dtsch. Chem. Ges.*, 8(2), 1056–1059.

Changeux, J.-P. and Connes, A. (1989). *Matière à pensée*. Odile Jacob, Paris.

Chartland, G. (2016). Highly irregular. In *Graph Theory, Favorite Conjectures and Open Problems*, Gera, R., Hedetniemi S., Larson C. (eds). Springer International Publishing, Switzerland.

Chudnovsky, M. (2014). The Erdös–Hajnal conjecture – a survey. *J. Graph Theory*, 75(2), 178–190. DOI: 10.48550/arXiv.1606.08827.

Chudnovsky, M. and Safra, S. (2008). The Erdös–Hajnal conjecture for bull-free graphs. *J. Comb. Theory Ser. B*, 98(6), 1301–1310.

Chudnovsky, M., Scott, A., Seymour, P., Spirkl, S. (2021). Erdös-Hajnal for graphs with no 5-hole. DOI: 10.48550/arXiv.2102.04994.

Corominas, E. (1984). On better-quasi-ordering countable trees. In *Proceedings of the Conference on Ordered Sets and their Applications*, Pouzet, M. and Richard, D. (eds). North-Holland, Amsterdam.

Courcier, J. (2008). Dominique Dubarle et la géométrie projective. *Revue des sciences philosophiques et théologiques*, 92(3), 623–636.

Coxeter, H.S. (1973). *Regular Polytopes*, 3rd edition. Dover Publications, New York.

Cuesta Dutari, N. (1959). *Matemática del orden*. Real Academia de Ciencias Exactas, Físicas y Naturales, Madrid.

Davey, B.A. and Priestley, H.A. (2002). *Introduction to Lattices and Orders*, 2nd edition. Cambridge University Press, Cambridge.

Dehmer, M. and Sivakumar, L. (2014). On comparability graphs: Theory and applications. In *Advances in Mathematica Chemistry and Applications*, volume 1, Basak, S.C., Restrepo, G., Villaveces, J.L. (eds). Bentham Science Publishers, Sharjah, U.A.E.

Dehornoy, P. (2003). Progrès récents sur l'hypothèse du continu. *Séminaire Bourbaki*, 55(915), 147–172.

Dehornoy, P. (2007). Au-delà du forcing: la notion de vérité essentielle en théorie des ensembles. In *Logique, dynamique et cognition*, Joinet, J.B. (ed.). Éditions de la Sorbonne, Paris.

Delahaye, P. (1994). Ensembles et hyperensembles. *Pour la Science*, 195, 93–97.

Delahaye, P. (2019). En finir avec l'hypothèse du continu. *Pour la Science*, 10(504), 26–36.

Deleuze, G. (2003). *Spinoza: Philosophie pratique*. Éditions de Minuit, Paris.

Deleuze, G. and Guattari, F. (1991). *Qu'est-ce que la philosophie?* Éditions de Minuit, Paris.

Dembowski, P. (1968). *Finite geometries*. Springer Science & Business Media, Berlin, Heidelberg, New York (reprint 2012).

Denys L'Aéropagite (1846). Traité de la Théologie mystique. In *Œuvres de Saint Denys L'Aéropagite*, translated by Mgr Darboy. Maison de la Bonne Presse, Paris.

Derrida, J. (1974). Le Parergon. *Digraphe*, 2, 3.

Derrida, J. (1978). *La Vérité en peinture*. Flammarion, Paris.

Derrida, J. and Owens, C. (1979). The Parergon. *JSTOR*, 9(Summer), 3–41.

Desanti, J.-T. (1980). La machine, le ruban. *Tel Quel*, 86, 18–25.

Descartes, R. (2016). *A Discourse on Method*, translated by Veitch, J. (Gutenberg-e-book, 2008) updated 2016.

DeWitt, R. (2010). *Worldviews: An Introduction to the History and Philosophy of Science*, 2nd edition. Wiley-Blackwell, West Sussex.

Diestel, R. (2001). Relating subsets of a poset, and a partition theorem for WQOs. *Order*, 18, 275–279.

Diestel, R. (2006). *Graph Theory*, 3rd edition. Springer, Berlin, Heidelberg New York.

Dieudonné, J. (1974). *Cours de géométrie algébrique*, 2 volumes. P.U.F, Paris.

Dieudonné, J. (1977). *Panorama des mathématiques pures*. Gauthier-Villars, Paris.

Dilworth, R.P. (1950). A decomposition theorem for partially ordered sets. *Ann. Math.*, 51(1), 161–166.

Dimopoulos, S., Gitman, V., Nielsen, D. S. (2021). The Virtual Large Cardinal Hierarchy, *arXiv:2109.06079v1*.

Dipert, R.R. (1997). The mathematical structure of the world: The world as graph. *J. Philos.*, 94(7), 329–358.

Dubarle, D. (1996). *L'Ontologie de Thomas d'Aquin*. Éditions du Cerf, Paris.

Dubarle, D. and Doz, A. (1972). *Logique et dialectique*. Larousse, Paris.

Duffus, D. and Rival, I. (1977). Path lengths in the covering graph. *Discrete Math.*, 19, 139–158.

Dushnik, B. and Miller, E.W. (1940). Concerning similarity transformations of linearly ordered sets. *Bull. Amer. Math. Soc.*, 46, 322–326.

Dushnik, B. and Miller, E.W. (1941). Partially ordered sets. *Am. J. Math.*, 63(3), 600–610.

Erdös, P. and De Bruijn, N.G. (1951). A colour problem for infinite graphs and a problem in the theory of relations. *Nederl. Akad. Wetensch. Indag. Math.*, 13, 311–313.

Erdös, P. and Hajnal, A. (1989). Ramsey-type theorems. *Discrete Appl. Math.*, 25(1–2), 37–52.

Erdös, P. and Rényi, A. (1963). Asymmetric graphs. *Acta Math. Hung.*, 14(3), 295–315.

Erdös, P. and Szekeres, G. (1935). A combinatorial problem in geometry. *Compos. Math.*, 2, 463–470.

Erné, M. (1982). *Einführung in die Ordnungstheorie*. Bibliographisches Institut, Mannheim.

Farrugia, A. (1999). Self-complementary graphs and generalisations: A comprehensive reference manual. Thesis, University of Malta.

Feng, Q., Magidor, M., Woodin, H. (1992). Universally Baire sets of reals. In *Set Theory of the Continuum*, Judah, H., Just, W., Woodin, H. (eds). Springer-Verlag, Inc., New York.

Ferguson, J. (1799). *Astronomy Explained Upon Sir Isaac Newtons Principles*, 10th edition. James Ferguson, London.

Filipov, N.D. (1980). Comparability graphs of partially ordered sets of different types. *Collq. Maths. Soc. Janos Bolyai*, 33, 373–380.

Finsler, P. (1926). Über die Grundlagen der Mengenlehre. I: Die Mengen und ihre Axiome. *Math. Z.*, 25, 683–713.

Finsler, P. (1963). Totalendliche Mengen. *Vierteljahresschrift der Naturforschenden Gesellschaft in Zürich*, 108, 141–152.

Fishburn, P.C. (1985). *Interval Orders and Interval Graphs: A Study of Partially Ordered Sets*. Wiley, New York.

Foucault, M. (1972). Mon corps, ce papier, ce feu. *Histoire de la Folie à l'âge classique*. Gallimard, Paris.

Fox, J. and Pach, J. (2012). String graphs and incomparability graphs. *Adv. Math.*, 230(3), 1381–1401.

van Fraassen, B. (1973). Values and the heart's command. *J. Philos.*, 70(1), 5–19.

Fraïssé, R. (1984). L'intervalle en théorie des relations, ses généralisations, filtre intervallaire et clôture d'une relation. In *Orders, Description and Roles*, Pouzet, M., Richard, D. (eds). North-Holland, Amsterdam.

Fraïssé, R. (1986). *Theory of Relations*. North-Holland, Amsterdam (2nd edition, Elsevier, 2000).

Fulkerson, D.R. (1956). Note on Dilworth's decomposition theorem for partially ordered sets. *Proc. Amer. Math. Soc.*, 7(4), 701–702.

Gallai, T. (1967). Transitiv Orientierbare Graphen. *Acta Math. Acad. Sci. Hungar.*, 18, 25–66.

Gallavotti, G. (2001). Quasi periodic motions from Hipparchus to Kolmogorov. *Rendiconti Accademia dei Lincei, Matematica e applicazioni*, 12, 125–152. See also: arXiv:chao-dyn/9907004v1.

Ganter, B. and Wille, R. (2003). Applied lattice theory: Formal concept analysis. In *General Lattice Theory*, 2nd edition, Grätzer, G. (ed.). Birkhäuser, Basel, Boston, Berlin.

Gardies, J.L. (1975). *La logique du temps*. P.U.F., Paris.

Gardner, M. (1992). *Fractal Music, Hypercards, and more Mathematical Recreations from Scientific American*. W. H. Freeman and Company, London.

Gauthier, Y. (1967). Logique hégélienne et formalisation. *Dialogue: Canadian Philosophical Review / Revue canadienne de philosophie*, 6(2), 151–165.

Gedenova, E. (1980). Lattices, whose covering graphs are s-graphs. *Colloq. Maths. Soc. Janos Bolyai*, 33, 407–435.

Gilmore, P.C. and Hoffman, A.J. (1964). A characterization of comparability graphs and of interval graphs. *Can. J. Math.*, 16, 539–548.

Gitman, V. (2007). Applications of the proper forcing axiom to models of Peano arithmetic. PhD Dissertation, Graduate Center of the City University of New York.

Godsil, C. and Royle, G. (2001). *Algebraic Graph Theory*. Springer-Verlag, New York.

Goldschmidt, V. (1953). *Le Système stoïcien et l'idée de temps*. Vrin, Paris.

Golumbic, M. (1980). *Algorithmic Graph Theory and Perfect Graphs*. Academic Press, New York.

Golumbic, M., Rotem, D., Urrutia, J. (1983). Comparability graphs and intersection graphs. *Discrete Math.*, 43, 37–46.

Gondran, M. and Minoux, M. (1979). *Graphes et Algorithmes*. Eyrolles, Paris.

Gramain, A. (1971). *Topologie des surfaces*. P.U.F., Paris.

Granger, G.-G. (1969). *Wittgenstein*. Seghers, Paris.

Granier, J. (1977). *Le Discours du Monde*. Le Seuil, Paris.

Guenther (1964). Das Problem einer Formalisierung der tranzendental-dialektischen logik. *Hegel-Studien*.

Gueroult, M. (1957). Logique, architectonique et structures constitutives des systèmes philosophiques. *Encyclopédie française*, XIX, 19-24-16.

Gueroult, M. (1979). *Philosophie de l'histoire de la philosophie*, volume 2. Dianoématique, Aubier-Montaigne, Paris.

Guillaumaud, J. (1980). Sauver la dialectique ? In *Science et Dialectique*, Vadée, M. (ed.). CNRS, Paris.

Gyárfás, A. (1997). Reflections on a problem of Erdös and Hajnal. In *The mathematics of Paul Erdös, II*, volume 14. Springer, Berlin, Heidelberg.

Hadot, P. (1968). *Porphyre et Victorinus*. Études Augustiniennes, Paris.

Hamkins, J.D. (2014). A multiverse perspective on the axiom of constructibility. *Infinity and truth*, Lect. Notes Ser. Inst. Math. Sci. Natl. Univ. Singap., 25, 25–45. DOI: 10.48550/arXiv.1210.6541.

Harary, F. (1964). On the reconstruction of a graph from a collection of subgraphs. In *Theory Graphs Appl.* (Proc. Sympos. Smolenice, 1963). Publ. House Czechoslovak Acad. Sci., Prague.

Harary, F. (1974). A survey of the reconstruction conjecture. In *A Survey of the Reconstruction Conjecture, Graphs and Combinatorics*, Bari, R.A., Harary, F. (eds), volume 406. Springer, Berlin, Heidelberg.

Harary, F. and Palmer, E. (1967). On the problem of reconstructing a tournament from sub-tournaments. *Monatsh. Math.*, 71, 14–23.

Harzheim, E. (2005). *Ordered Sets*, volume 7. Springer New York, NY.

Hausdorff, F. (1949). *Grundzuege Der Mengenlehre* (1914). Chelsea Publishing Company, New York.

He, Y. and Schnabel, M.A. (2018). The Logic behind Lattice Windows in the Lingering Garden: Parametric Generation of Chinese Abstract Indigenous Patterns. In *Learning, Prototyping and Adapting, Short Paper Proceedings of the 23rd International Conference on Computer-Aided Architectural Design Research in Asia (CAADRIA)*, Huang, W., Williams, M., Luo, D., Wu, Y., Lin, Y. (eds.).

Heesch, H. (1969). *Untersuchungen zum Vierfarbenproblem*. Bibliographisches Institut, Mannheim.

Heidegger, M. (2021). *The Metaphysics of German Idealism: A New Interpretation of Schelling's Philosophical Investigations into the Essence of Human Freedom and Matters*, translated by Moore, I.A. and Therezo, R.. Wiley, New York, NY.

Herbert, J. and Varenne, J. (1985). *Vocabulary of Hinduism*. Dervy, Paris.

Higman, G. (1952). Ordering by divisibility in abstract algebras. *Proc. London Math. Soc.*, 2, 326–336.

Hilbert, D. (1993). *Theory of Algebraic Invariants*. Cambridge University Press, Cambridge.

Hirschfeld, J.W.P. (1979). *Projective Geometries over Finite Field*. Clarendon Press, Oxford.

Hobbes, T. (1651). *Leviathan or the Matter, Forme, & Power of a Common-wealth Ecclesiastical and Civil*, printed for Andrew Crooke, at the Green Dragon in St. Paul's Churchyard, Malmesbury.

Holton, D.A. and Sheehan, J. (1993). *The Petersen Graph*. Cambridge University Press, Cambridge.

Houellebecq, M. (2022). *Anéantir*. Flammarion, Paris.

Jarry, A. (2007). *Les minutes de sable mémorial* (1894), followed by *César Antéchrist*. Grasset, Paris.

Jaspers, K. (1951). *Way to Wisdom: An Introduction to Philosophy*, translated by Manheim, R. Owsley, R.M.. Yale University Press, New Haven, CT.

Jech, T. (2003). *Set Theory*. Springer Verlag, Berlin, Heidelberg, New York.

Kanamori, A. (2005). *The Higher Infinite, Large Cardinals in Set Theory from their Beginning*. Springer Verlag, Berlin, Heidelberg, New York.

Kant, E. (1922). *Critique of Pure Reason*. The MacMillan Company, McMillan & Co, London.

Kaufmann, A. (1973). *Introduction à la théorie des sous-ensembles flous, à l'usage des ingénieurs*, volume 1. Éléments théoriques de base, Masson, Paris.

Kaufmann, A. (1975). *Introduction à la théorie des sous-ensembles flous, à l'usage des ingénieurs*, volume 2. Masson, Paris.

Kaufmann, A. and Boulaye, G. (1978). *Théorie des treillis en vue des applications*. Masson, Paris.

Kelly, D. (1985). Comparability graphs. In *Graphs and Orders*, Rival, I. (ed.). D. Reidel Publishing Company, Dordrecht.

Kocay, W.L. (1987). A family of non-reconstructible hypergraphs. *J. Combin. Theory Ser. B*, 42, 46–63.

König D. (1990). *Theorie der endlichen und unendelichen Graphen* (*Theory of finite and infinite graphs*), translated by McCoart R. Birkhäuser Verlag, Basel.

Korzybski, A. (1933). *Korzybski, Science and Sanity. An Introduction to Non-Aristotelian Systems and General Semantics*. The International Non-Aristotelian Library Pub. Co, New York.

Kosok, M. (1966). The formalization of Hegel's dialectical logic: Its formal structure, logical interpretation and intuitive foundation. *Int. Philos. Quart.*, 6(4), 596-63.

Kosok, M. (1972). The formalization of Hegel's dialectical logic: Its formal structure, logical interpretation and intuitive foundation. In *Hegel: A Collection of Critical Essays*, MacIntyre, A. (ed.). University of Notre Dame Press, Notre Dame, IN.

Koyré, A. (1929). *La philosophie de Jacob Boehme, étude sur les origines de la métaphysique allemande*. Vrin, Paris (reprint 2002).

Kruja, E., Marks, J. Blair, A., Waters, R. (2002). A short note on the history of graph drawing. In *Graph Drawing*, Mutzel, P., Œnger, M.J., Leipert, S. (eds). Springer, Berlin, Heidelberg.

Kruskal, J.B. (1960). Well-quasi-ordering, the tree theorem and V'azsonyi's conjecture. *Trans. Amer. Math. Soc.*, 95, 210–225.

Kurepa, G. (1935). Ensembles ordonnés et ramifiés. Thesis, Paris.

Ladrière, J. (1972). Système (épistémologie). *Encyclopaedia Universalis*, Paris.

Lakatos, I. (1976). *Proofs and Refutations*. Cambridge University Press, Cambridge.

Larson, C.E. (2011). The critical independence number and an independence decomposition. *Eur. J. Comb.*, 32, 294–300.

Lask, E. (1911). *Die Logik der Philosophie und die Kategorien-lehre*, J.C.B. Mohr (P. Siebeck), Tubingen.

Laver, R. (1971). On Fraïssé's order type conjecture. *Ann. Math.*, 93, 89–111.

Lemmon, E.J. (1962). Moral dilemmas. *Philos. Rev.*, 71(2), 139–158.

Levi, F.W. (1942). *Finite Geometrical Systems*. University of Calcutta, Calcutta.

Levi-Strauss, C. (1950). Introduction l'oeuvre de Marcel Mauss. In *Sociologie et Anthropologie*, Mauss, M. (ed.). P.U.F, Paris.

Levi-Strauss, C. (1962). *La Pensée sauvage*. Plon, Paris.

Levit, V.E. and Mandrescu, E. (2013). On the structure of the minimum critical independent set of a graph. *Discrete Math.*, 313, 605–610.

L'Huillier, S.-A.-J. (1812-13). Mémoire sur la polyédrométrie. *Ann. Math.*, 3, 169–189.

Louis, A.A., Johnston, I.G., Dingle, K., Greenbury, S.M., Camargo, C.Q., Doye, J.P.K., Ahnert, S.E., (2022). Symmetry and simplicity spontaneously emerge from the algorithmic nature of evolution. *bioRxiv* (preprint, PNAS).

Lovasz, L. (1983). Perfect graphs. In *Selected Topics in Graph Theory*, volume 2. Academic Press, London.

MacNeille, H.M. (1937). Partially ordered sets. *Trans. Bitter. Math. Soc.*, 42, 416–460.

Maddy, P. (1988a). Believing the axioms I. *The Journal of Symbolic Logic*, 53(2), 481–511.

Maddy, P. (1988b). Believing the axioms II. *The Journal of Symbolic Logic*, 53(3), 736–764.

Maddy, P. (1997). *Naturalism in Mathematics*. Clarendon Press, Oxford.

Maddy, P. (2011). *Defending the Axioms: On the Philosophical Foundations of Set Theory*. Oxford University Press, Oxford.

Makinson, D. C. (1965). The paradox of the preface. *Analysis*, 25(6), 205–207.

Malebranche, N. (1958). *Traité de la Nature et de la Grâce* (1684). Vrin, Paris.

Mannheim, K. (1964). Die Strukturanalyse der Erkenntnistheorie. Berlin, Reuther und Reichard, *Kant Studien / Ergänzungsheft* volume 57, 1922. (Reprint in *Wissenssoziologie. Auswahl aus dem Werk*, Wolff K.H. (ed.). Berlin-Neuwied, Luchterhand.)

Marconi, D. (1979). *La Formalizzazione delle Dialettica*. Rosenberg and Sellier, Turin.

Markov, A. (1958). Insolubility of the problem of homeomorphy. *Proc. Intern. Congress of Math.*, 300–306. Cambridge University Press, Cambridge.

McKay, B.D. (1997). Small graphs are reconstructible. *Australas. J. Combin.*, 15, 123–126.

Mélès, B. (2016). *Les classifications des systèmes philosophiques*. Vrin, Paris.

Milner, E.C. (1985). Basic wqo and bqo theory. In *Graphs and Orders*, Rival, I. (ed.). D. Reidel Publishing Company, The Netherlands.

Monjardet, B. (1968). Géométries planes finies. Combinatoire et structures algébriques. *Mathematiques et sciences humaines*, 21, 39–51.

Moorhouse, G.E. (2006). On projective planes of order less than 32. In *Finite Geometries, Groups, and Computation*, Hulpke, A. et al. (eds). de Gruyter, Berlin.

Moorhouse, G.E. (2020). Projective planes, finite and infinite. Waterloo C & O Seminar, 7 September 2020 [Online]. Available at: https://homepages.dcc.ufmg.br › uploads › 2020/09.

Nadis, S. (2021). New proof reveals that graphs with no pentagons are fundamentally different. *Quanta Magazine* [Online]. Available at: https://www.quantamagazine.org/new-proof-reveals-that-graphs-with-no-pentagons-are-fundamentally-different-20210426/. [Accessed 26 April 2021].

Nash-Williams, C.St.J.A. (1967). Infinite graph. *J. Comb. Theory*, 3, 286–301.

Nash-Williams, C.St.J.A. (1968). On better-quasi-ordering transfinite sequences. *Proc. Camb. Phil. Soc.*, 64, 273–290.

Nash-Williams, C.St.J.A. (1978). The reconstruction problem. *Selected Topics in Graph Theory*, 1, 205–236.

Nielsen, D.S. (2018). Diagrams. *A Travel Guide to the Large Cardinals* (blog) [Online]. Available at: https://dansnielsen.wordpress.com.

Nimbhorkar, S.K., Wasadikar, M.P., Pawar, M.M. (2010). Coloring of lattices. *Math. Slovaca*, 60, 419–434.

Oderberg, D.S. (2011). The world is not an asymmetric graph. *Analysis*, 71, 3–10.

Oderberg, D.S. (2012). No potency without actuality: The case of graph theory. In *Contemporary Aristotelian Metaphysics*, Tahko, T.E. (ed.). Cambridge University Press, New York.

Olariu, S. (1992). On sources of comparability graphs with applications. *Discrete Math.*, 110(1–3), 289–292.

Ore, O. (1962). *Theory of Graphs*. American Mathematical Society, Providence, RI.

Parrochia, D. (1990). Formalisation boolénne et intuitionniste de la logique hégélienne. *Acta Universitatis Lodziensis, Folia philosophica*, 7, 111–123.

Parrochia, D. (1991). *Mathematics & Existence*. Champ Vallon, Seyssel.

Parrochia, D. (1993a). *La Raison systématique*. Vrin, Paris.

Parrochia, D. (1993b). *Philosophie des réseaux*. P.U.F., Paris.

Parrochia, D. (1997). *Les grandes théories scientifiques du XXe siècle*. P.U.F, Paris.

Parrochia, D. (1999). Afterword to: Paul Finsler. In *De la vie après la mort: mathematiques et metaphysique*, translated by Marine, E.

Parrochia, D. (2004). *Multiples*. Le Corridor Bleu, Paris.

Parrochia, D. (2006). *Philosophie et musique contemporaine*. Champ Vallon, Seyssel.

Parrochia, D. (2012). Quasi-orders, networks, probability spaces and deterministic chaos, weak forms of open rationality. In *Chinese & French Views on Knowledge and Society Today*, Rieu, A.-M. (ed.). Éditions des Archives contemporaines, Paris.

Parrochia, D. (2018). *Mathematics and Philosophy*. ISTE Ltd, London and Wiley, New York.

Parrochia, D. (2020). On the number of edges of comparability and incomparability graphs, hal-02972652.

Parrochia, D. (2021). On von Weizsäcker's philosophy of Quantum Mechanics. *arXiv:2103.07311v1*.

Parrochia, D. and Neuville, P. (2013). *Towards a General Theory of Classifications*. Birkäuser, Basel.

Perles, M.A. (1963). On Dilworth's theorem in the infinite case. *Isr. J. Math.*, 1(2), 108–109.

Perlmutter, N.J. (2013). The large cardinals between supercompact and almost-huge. *arXiv:1307.7387v1*.

Petitot, J. (2009). A transcendental view on the continuum: Woodin's conditional platonism. *Intellectica*, 51, 93–133.

Pierce, C.S. (1885). On the algebra of logic: A contribution to the philosophy of notation. *Am. J. Math.*, 7(2), 180–196.

Pierce, R.S. (1970). Classification problems. *Math. Syst. Theory*, 4(1), 65–80.

Ponasse, D. and Carréga, J.C. (1979). *Algèbre et topologie booléennes*. Masson, Paris.

Popper, K.R. (1940). What is dialectic? *Mind*, 49(196), 403–426. (Reprinted in *Conjectures and Refutations: The Growth of Scientific Knowledge*. Basic Books, New York, 1962; Routledge Classics, London, 2002.)

Pouzet, M. (1978). Ensemble ordonné universel recouvert par deux chaînes. *Journal of Combinatorial Theory, Ser. B*, 25, 1–25.

Pouzet, M. (1985). Applications of well-quasi-ordering and better-quasi-ordering. In *Graphs and Orders*, Rival, I. (ed.). D. Reidel Publishing Company, 503–519.

Queneau, R. (1961). *Cent mille milliards de poèmes*. Gallimard, Paris.

Quine, W.V.O. (1969). *Ontological Relativity und Other Essays in Philosophy*. Columbia University Press, New York.

Rado, R. (1949). Axiomatic treatment of rank in infinite sets. *Can. J. Math.*, 1, 337–343.

Rado, R. (1954). Partial well-ordering of sets of vectors. *Mathematika*, 1, 89–95.

Rao, S.B. (1977). Cycles in self-complementary graphs. *J. Comb. Th.*, 22(1), 1–9.

Rasiowa, H. (1974). *An Algebraic Approach to Non-classical Logics*. North-Holland, Amsterdam.

Rhodes, J. and Silva, P.C. (2015). *Boolean Representations of Simplicial Complexes and Matroids*. Springer Verlag, Cham, Heidelberg, New York, Dordrecht, London.

Rittberg, C.J. (2015). How Woodin changed his mind: New thoughts on the Continuum Hypothesis. *Archive for History of Exact Sciences*, 69(2), 125–151.

Robertson, N., Sanders, D., Seymour, P., Thomas, R. (1997). The four color theorem. *J. Comb. Theory, Ser. B*, 70, 2–44.

Rogowski, L.S. (1964). *Logika Kierunkowa A Heglowska Teza O Sprzecznosci Zmiany*. Torun.

Russo, L. (2004). *The Forgotten Revolution: How Science Was Born in 300 BC and Why It Had to Be Reborn*. Springer, Berlin,

Sallantin, J. (1986). Logique et comportement des systèmes rationnels, une esquisse d'épistémologie. *Intelligence des mécanismes, mécanismes de l'intelligence*, Fondation Diderot, Fayard, Paris.

Schiaparelli, G. (1926). Scritti sulla storia dell' astronomia antica. Part I. Tomo II. In *Le sfere omocentriche di Eudosso, di Callippo e di Aristotele*. Zanichelli, Bologna.

Scholem, G. (1949). The Curious History of the Six-Pointed Star: How the "Magen David" became the Jewish Symbol. *Commentary Magazine*.

Schröder, E. (1890). *Vorlesungen über die Algebra der Logik*, 3 volumes. B.G. Teubner, Leipzig 1890-1905 (Reprinted Thoemmes Press, Chelsea, 2000).

Schröder, B.S.W. (2002). *Ordered Sets, An Introduction*. Birkhäuser, Boston, Basel, Berlin.

Serres, M. (1968). *Le système de Leibniz et ses modèles mathématiques*, 2 volumes. P.U.F., Paris.

Shackel, N. (2011). The world as a graph: Defending metaphysical graphical structuralism. *Analysis*, 71(1), 10–21.

Smarandache, F. (2002). Neutrosophy, a new branch of philosophy. *Multiple Valued Logic/Int. J.*, 8(3), 297–384.

Smarandache, F. and Broumi, S. (2020). *Neutrosophic Graph Theory and Algorithms*. IGI Global, Hershey PA.

Sommerville, D.M.Y. (1929). *An Introduction to Geometry in n Dimensions*. Methuen, London.

Stanley, R. (1970). Structure of incidence algebras and their automorphism groups. *Bull. Amer. Math. Soc.*, 76, 1236–1239.

Stanley, R. (1988). Differential posets. *J. Am. Math. Soc.*, 1, 919–961.

Stanley, R. (2004). *Enumerative Combinatorics*, volume 1. China Machine Press.

Stirner, M. (2017). *The Unique and Its Property*. Underworld Amusements, Baltimore.

Stocker, M. (1971). 'Ought' and 'Can'. *Australas. J. Philos.*, 49, 303–316.

Stockmeyer, P.K. (1977). The falsity of the reconstruction conjecture for tournaments. *J. Graph Theory*, 1, 19–25.

Stockmeyer, P.K. (1981). A census of non-reconstructable digraphs, I: Six related families. *J. Combin. Theory Ser. B*, 31, 232–239.

Sylvester, J.J. (1878). Chemistry and algebra. *Nature*, 17, 284.

Takashi, I. (1998). Professor Quine on Japanese classifiers. *Ann. Japan Assoc. Philos. Sci.*, 9(3), 111–118.

Todeschini, R., Consonni, V., Mannhold, R. (2002). *Handbook of Molecular Description*. Wiley-WCH, Weinheim.

Trendelenburg, F.A. (1840). *Logische Untersuchungen*, G. Bethge, Berlin.

Trotter, W.T. (1995). Partially ordered sets. In *Handbook of Combinatorics*, Graham, R.L., Grötschel, M., Lovász, L. (eds). Elsevier, Amsterdam.

Vannier, M.-A. (1998). Aux sources de la voie négative. *Revue des Sciences Religieuses*, 72-4, 403–419.

Vasantha Kandasamy, W.B., Ilanthenral, K., Smarandache, F. (2015). *Neutrosophic Graphs: A New Dimension to Graph Theory*. EuropaNova, ASBL, Brussels.

Vuillemin, J. (1984). *Nécessité et Contingence, l'aporie de Diodore et les systèmes philosophiques*. Minuit, Paris.

Wachs, M.L. (2004). *Poset Topology: Tools and Applications*. Lecture notes IAS/Park City Graduate Summer School in Geometric Combinatorics.

Wagner, K. (1937). Über eine Eigenschaft der ebenen Komplexe. *Mathematische Annalen*, 114(1), 570–590.

Wasadikar, M.P. and Survase, P.A. (2012). Incomparability graphs of lattices. In *International Conference on Mathematical Modelling and Scientific Computation – Mathematical Modelling and Scientific Computation*, ICMMSC 2012, Balasubramaniam, P. and Uthayakumar, R. (eds). Springer, Heidelberg.

Weil, E. (1967). *Logique de la Philosophie*. Vrin, Paris.

von Weizsäcker, C.F. (1971). The unity of physics. In *Quantum Theory and Beyond*, Bastin, T. (ed.). Cambridge University Press, London, New York.

Williams, B.A.O. (1966). Consistency and realism. *Proc. Aristot. Soc*, Supplementary Volumes, 40(1,10), 1–22.

Williams, R. (1979). *The Geometrical Foundation of Natural Structure: A Source Book of Design*. Dover, New York.

Williams, B.A.O and Atkinson, W.F. (1965). Ethical consistency. *Proc. Aristot. Soc*, Supplementary Volumes, 39, 103–138.

Woodin, W.H. (2001a). The continuum hypothesis I. *Not. AMS*, 48(6), 567–576.

Woodin, W.H. (2001b). The continuum hypothesis II. *Not. AMS*, 48(7), 681–690.

Wu, Y. and Shizen, Z. (2012). Incidence matrix and cover matrix of nested interval orders. *Electron. J. Lin. Algebra*, 23, 43–65.

Yang, Y. (1988). The reconstruction conjecture is true if all 2-connected graphs are reconstructible. *J. Graph Theory*, 12, 237–243.

Zhu, X, (2004). Perfect graphs for generalized colouring-circular perfect graphs. In *Graphs, Morphisms and Statistical Physics*, volume 63. Amer. Math. Soc., Providence, RI.

Index

A, B

C

D, E, F

G

H, L

M, N

O, P

Q, S

T, V

W, Y

Other titles from

ISTE

in

Mathematics and Statistics

2022

CHAKRAVARTHY Srinivas R.
Introduction to Matrix-Analytic Methods in Queues 1: Analytical and Simulation Approach – Basics
Introduction to Matrix-Analytic Methods in Queues 2: Analytical and Simulation Approach – Queues and Simulation

DE SAPORTA Benoîte, ZILI Mounir
Martingales and Financial Mathematics in Discrete Time

LESFARI Ahmed
Integrable Systems

RADCHENKI Vadym M.
General Stochastic Measures: Integration, Path Properties and Equations

SIMON Jacques
Distributions
(Analysis for PDEs Set – Volume 3)

2021

KOROLIOUK Dmitri, SAMOILENKO Igor
Random Evolutionary Systems: Asymptotic Properties and Large Deviations

MOKLYACHUK Mikhail
Convex Optimization: Introductory Course

POGORUI Anatoliy, SWISHCHUK Anatoliy, RODRÍGUEZ-DAGNINO Ramón M.
Random Motions in Markov and Semi-Markov Random Environments 1: Homogeneous Random Motions and their Applications
Random Motions in Markov and Semi-Markov Random Environments 2: High-dimensional Random Motions and Financial Applications

PROVENZI Edoardo
From Euclidean to Hilbert Spaces: Introduction to Functional Analysis and its Applications

2020

BARBU Vlad Stefan, VERGNE Nicolas
Statistical Topics and Stochastic Models for Dependent Data with Applications

CHABANYUK Yaroslav, NIKITIN Anatolii, KHIMKA Uliana
Asymptotic Analyses for Complex Evolutionary Systems with Markov and Semi-Markov Switching Using Approximation Schemes

KOROLIOUK Dmitri
Dynamics of Statistical Experiments

MANOU-ABI Solym Mawaki, DABO-NIANG Sophie, SALONE Jean-Jacques
Mathematical Modeling of Random and Deterministic Phenomena

2019

BANNA Oksana, MISHURA Yuliya, RALCHENKO Kostiantyn, SHKLYAR Sergiy
Fractional Brownian Motion: Approximations and Projections

GANA Kamel, BROC Guillaume
Structural Equation Modeling with lavaan

KUKUSH Alexander
Gaussian Measures in Hilbert Space: Construction and Properties

LUZ Maksym, MOKLYACHUK Mikhail
Estimation of Stochastic Processes with Stationary Increments and Cointegrated Sequences

MICHELITSCH Thomas, PÉREZ RIASCOS Alejandro, COLLET Bernard, NOWAKOWSKI Andrzej, NICOLLEAU Franck
Fractional Dynamics on Networks and Lattices

VOTSI Irene, LIMNIOS Nikolaos, PAPADIMITRIOU Eleftheria, TSAKLIDIS George
Earthquake Statistical Analysis through Multi-state Modeling
(Statistical Methods for Earthquakes Set – Volume 2)

2018

AZAÏS Romain, BOUGUET Florian
Statistical Inference for Piecewise-deterministic Markov Processes

IBRAHIMI Mohammed
Mergers & Acquisitions: Theory, Strategy, Finance

PARROCHIA Daniel
Mathematics and Philosophy

2017

CARONI Chysseis
First Hitting Time Regression Models: Lifetime Data Analysis Based on Underlying Stochastic Processes
(Mathematical Models and Methods in Reliability Set – Volume 4)

CELANT Giorgio, BRONIATOWSKI Michel
Interpolation and Extrapolation Optimal Designs 2: Finite Dimensional General Models

CONSOLE Rodolfo, MURRU Maura, FALCONE Giuseppe
Earthquake Occurrence: Short- and Long-term Models and their Validation
(Statistical Methods for Earthquakes Set – Volume 1)

D'AMICO Guglielmo, DI BIASE Giuseppe, JANSSEN Jacques, MANCA Raimondo
Semi-Markov Migration Models for Credit Risk
(Stochastic Models for Insurance Set – Volume 1)

GONZÁLEZ VELASCO Miguel, del PUERTO GARCÍA Inés, YANEV George P.
Controlled Branching Processes
(Branching Processes, Branching Random Walks and Branching Particle Fields Set – Volume 2)

HARLAMOV Boris
Stochastic Analysis of Risk and Management
(Stochastic Models in Survival Analysis and Reliability Set – Volume 2)

KERSTING Götz, VATUTIN Vladimir
Discrete Time Branching Processes in Random Environment
(Branching Processes, Branching Random Walks and Branching Particle Fields Set – Volume 1)

MISHURA YULIYA, SHEVCHENKO Georgiy
Theory and Statistical Applications of Stochastic Processes

NIKULIN Mikhail, CHIMITOVA Ekaterina
Chi-squared Goodness-of-fit Tests for Censored Data
(Stochastic Models in Survival Analysis and Reliability Set – Volume 3)

SIMON Jacques
Banach, Fréchet, Hilbert and Neumann Spaces
(Analysis for PDEs Set – Volume 1)

2016

CELANT Giorgio, BRONIATOWSKI Michel
Interpolation and Extrapolation Optimal Designs 1: Polynomial Regression and Approximation Theory

CHIASSERINI Carla Fabiana, GRIBAUDO Marco, MANINI Daniele
Analytical Modeling of Wireless Communication Systems
(Stochastic Models in Computer Science and Telecommunication Networks Set – Volume 1)

GOUDON Thierry
Mathematics for Modeling and Scientific Computing

KAHLE Waltraud, MERCIER Sophie, PAROISSIN Christian
Degradation Processes in Reliability
(Mathematial Models and Methods in Reliability Set – Volume 3)

KERN Michel
Numerical Methods for Inverse Problems

RYKOV Vladimir
Reliability of Engineering Systems and Technological Risks
(Stochastic Models in Survival Analysis and Reliability Set – Volume 1)

2015

DE SAPORTA Benoîte, DUFOUR François, ZHANG Huilong
Numerical Methods for Simulation and Optimization of Piecewise Deterministic Markov Processes

DEVOLDER Pierre, JANSSEN Jacques, MANCA Raimondo
Basic Stochastic Processes

LE GAT Yves
Recurrent Event Modeling Based on the Yule Process
(Mathematical Models and Methods in Reliability Set – Volume 2)

2014

COOKE Roger M., NIEBOER Daan, MISIEWICZ Jolanta
Fat-tailed Distributions: Data, Diagnostics and Dependence
(Mathematical Models and Methods in Reliability Set – Volume 1)

MACKEVIČIUS Vigirdas
Integral and Measure: From Rather Simple to Rather Complex

PASCHOS Vangelis Th
Combinatorial Optimization – 3-volume series – 2nd edition
Concepts of Combinatorial Optimization / Concepts and Fundamentals – volume 1
Paradigms of Combinatorial Optimization – volume 2
Applications of Combinatorial Optimization – volume 3

2013

COUALLIER Vincent, GERVILLE-RÉACHE Léo, HUBER Catherine, LIMNIOS Nikolaos, MESBAH Mounir
Statistical Models and Methods for Reliability and Survival Analysis

JANSSEN Jacques, MANCA Oronzio, MANCA Raimondo
Applied Diffusion Processes from Engineering to Finance

SERICOLA Bruno
Markov Chains: Theory, Algorithms and Applications

2012

BOSQ Denis
Mathematical Statistics and Stochastic Processes

CHRISTENSEN Karl Bang, KREINER Svend, MESBAH Mounir
Rasch Models in Health

DEVOLDER Pierre, JANSSEN Jacques, MANCA Raimondo
Stochastic Methods for Pension Funds

2011

MACKEVIČIUS Vigirdas
Introduction to Stochastic Analysis: Integrals and Differential Equations

MAHJOUB Ridha

Recent Progress in Combinatorial Optimization – ISCO2010

RAYNAUD Hervé, ARROW Kenneth
Managerial Logic

2010

BAGDONAVIČIUS Vilijandas, KRUOPIS Julius, NIKULIN Mikhail
Nonparametric Tests for Censored Data

BAGDONAVIČIUS Vilijandas, KRUOPIS Julius, NIKULIN Mikhail
Nonparametric Tests for Complete Data

IOSIFESCU Marius *et al.*
Introduction to Stochastic Models

VASSILIOU PCG
Discrete-time Asset Pricing Models in Applied Stochastic Finance

2008

ANISIMOV Vladimir
Switching Processes in Queuing Models

FICHE Georges, HÉBUTERNE Gérard
Mathematics for Engineers

HUBER Catherine, LIMNIOS Nikolaos *et al.*
Mathematical Methods in Survival Analysis, Reliability and Quality of Life

JANSSEN Jacques, MANCA Raimondo, VOLPE Ernesto
Mathematical Finance

2007

HARLAMOV Boris
Continuous Semi-Markov Processes

2006

CLERC Maurice
Particle Swarm Optimization